THE
INVADERS

THE
INVADERS

How Humans and Their Dogs Drove Neanderthals to Extinction

PAT SHIPMAN

The Belknap Press of
Harvard University Press
Cambridge, Massachusetts
London, England
2015

Library of Congress Cataloging-in-Publication Data
Shipman, Pat, 1949–
The invaders : how humans and their dogs drove Neanderthals
to extinction / Pat Shipman.
      pages cm
Includes bibliographical references and index.
ISBN 978-0-674-73676-4 (hardcover : alk. paper)
  1. Human beings—Origin.   2. Human beings—Migrations.
3. Human evolution.   4. Neanderthals.   5. Dogs—Evolution.
6. Human-animal relationships—History.   7. Predation
(Biology)   I. Title.
  GN286.S55 2015
  569.9—dc23
                                                      2014034978

TO ZELDA

for teaching me about predator–prey relationships

AND FOR THE WOLF KNOWN AS *06*

for teaching me what being a wolf means

CONTENTS

PREFACE

This book draws together many disparate ideas from different fields of study and personal experiences that have enlightened me. One of the most important of these ideas began forming in 2009, when I read a remarkable paper by Mietje Germonpré and a group of her colleagues reporting that they had developed a way to distinguish dogs from wolves through statistical analyses of their crania. To their surprise, and mine, using this impeccable technique, they were able to identify an early dog skull that turned out to be about 32,000 years old—about 18,000 years earlier than the domestication of dogs had previously been believed to occur. After reading and rereading this article carefully, I was convinced that the researchers were right. I began pondering what having domestic dogs would mean to our early modern human ancestors and whether this event might be related to the extinction of Neanderthals, then generally accepted to have occurred just a few thousand years more recently than dogs appeared, according to Germonpré's work. I read extensively about fascinating work being done on the biological, behavioral, and genetic differences between dogs and wolves.

Another strand in the web of hypothesis testing occurred because I am fortunate enough to own a small holiday home on the Caribbean island of Little Cayman. Little Cayman's main and practically only claim to fame revolves around its superb beaches and unusually pristine coral reef system. It is a mecca for scuba divers and snorkelers who are happy to deal with the small population (about 175 permanent residents), its limited area (one mile by ten miles), its scarcity of shops and stores, and its focus on ecotourism. My husband and I love it for precisely those reasons. But in 2008, the lionfish (*Pterois volitans*) began to turn up on our reefs for the first time. The fish is beautiful, but it belongs in the Indo-Pacific. A deadly predator that eats other fish and crustaceans voraciously, lionfish also carry poison in their elegant spines—poison that is nasty enough to be a danger to divers who encounter it. Lionfish reproduce so rapidly that they can decimate those fishy and crustacean inhabitants of Caribbean coral reefs which do not recognize them as predators.

To protect the natural beauty and tourism industry of Little Cayman, something needed to be done about this invasive species. Little Cayman began an all-volunteer effort, holding a weekly cull to catch and remove lionfish. The resorts split the cost of the effort; a team of dive masters willing to undertake the effort was assembled; chefs at the resorts began to plan meals based on lionfish; fund-raisers were held; and data on the size, density, and diet of lionfish were compiled (for example, see T. K. Frazer, C. A. Jacobi, M. A. Edwards, S. C. Barry, and C. M. Manfrino, "Coping with the Lionfish Invasion: Can Targeted Removals Yield Beneficial Effects?" *Reviews in Fisheries Science* 20 (2012): 185–191). Inspired by this very real problem, I began to educate myself about invasive species and the developing field of invasion biology.

Though my interest was spurred by marine biology, I soon realized that the classic anthropological question of why Neanderthals went extinct, when early modern humans did not, could be analyzed fruitfully from the perspective of invasion biology. This offered some new insights on an old problem.

As I probed deeper into invasion biology, I read more and more about the reintroduction of gray wolves into Yellowstone National Park. Strictly speaking, this case was not a natural invasion but a manufactured one, designed to restore the Yellowstone ecosystem to its natural state before settlers, trappers, and ranchers moved into the area and exterminated wolves. However, because this was a long-anticipated and well-planned restoration, there were excellent and abundant data on the other, less-feared mammals as well as the plants and birds and their distribution throughout the park and its surrounds. The process of gray wolf reintroduction in 1995 and 1996 and its effects on many aspects of the Greater Yellowstone Ecosystem were counted, compiled, watched, filmed, documented, and analyzed. It was a superb analogue, I felt, for the entry of predatory modern humans into the Eurasian ecosystem, which happened perhaps 45,000 years ago.

In 2012, an old friend of mine, Mary Stapleton, invited me to join her on a remarkable course being given in Yellowstone called "The Stewardship of Public Lands: Politics and the Yellowstone Ecosystem." Organized by the American Association of State Colleges and Universities, the course was attended by faculty or administrators in higher education. We were taken through Yellowstone by expert wildlife guides and educated by people on all sides of various debates, from ranchers and others who lived on land surrounding the park, to park rangers, ecologists, animal behaviorists, and managers. Yellowstone is one of the world's most extraordinary places, so seeing it is always a thrill. But seeing it and learning the perspectives of these various stakeholders was eye-opening for me. By seeing what was going on, hearing it, and living it for a short while, I was able to catch a stunning glimpse of an ecosystem confronted with a dangerous top predator.

In 2013, I attended a marvelous conference in Kraków that gave me great deal of information about sites that had yielded thousands of mammoth bones and enabled me to visit some of the more spectacular ones. I also contacted friends new and old at the cutting edge of research on Neanderthals, early modern humans, genetics, archaeology, dating techniques, dogs, wolves, mammoths, animal behavior, and numerous other fields. I cannot thank those individuals enough who educated me, argued or agreed with me, sent me unpublished or newly published materials, let me use their illustrations, and helped me synthesize disparate types of data into a whole. I would like to thank Savannah Barry, Ofer Bar-Yosef, Hervé Bocherens, George Chaplin, Steve Churchill, Silvana Condemi, Nick Conard, Katerina Douka, Dorothée Drucker, Holly Dunsworth, Clive Finlayson, Dan Fisher, Jennifer French, Mietje Germonpré, Brian Hare, Christine Hertler, Tom Higham, Jeff Hoffecker, Nina Jablonski, Martina Lázničková-Galetová, Jeff Mathison (mapmaker extraordinaire), George Mehaffy, Paul Mellars, Dick Mol, Susanne Münzel, Neil van Niekirk, Lorenzo Rook, Chris Ruff, Beth Shapiro, John Shea, Doug Smith, Mary Stapleton, Olga Soffer, Jiří Svoboda, Olaf Thalmann, Sarah Tischkoff, Erik Trinkaus, Blaire Van Valkenburgh, Rebekkah Volmer, Bob

Wayne, Piotr Wojtal, and Jaroslaw Wylczyński for their generosity and help. If I have forgotten someone, it is due to an imperfect memory, not lack of gratitude.

Themes and ideas presented in Chapter 7 were developed in an earlier article I published in 2012, "The Cost of the Wild," in *American Scientist* 100:254–257. Similarly, Chapter 14 is informed by an article I published in 2012, "Do the Eyes Have It?" in *American Scientist* 100: 198–201.

I also thank my editors, Elizabeth Knoll and Michael Fisher; my agent, Michelle Tessler; and, as always, my husband, for assistance and inspiration.

THE
INVADERS

AND HE IS US

Most books about science do not open by declaring that some of the most revered scientists in the world have made a big mistake, but here goes: They're wrong. Although it is almost universally acknowledged in the scientific community that invasive species are a terrible problem—altering ecosystems, causing extinctions, lessening biodiversity—there is a glaring omission in our catalog of invasive species. You can check for yourself by going to the website http://en.wikipedia.org/wiki/List_of_globally_invasive_species, where you will find a list of the 100 worst invasive species, drawn from the Global Invasive Species Database. That list is maintained by the Invasive Species Specialist Group of the International Union for the Conservation of Nature (IUCN). In other words, some of the most knowledgeable, clever, concerned, and well-informed people in the world are worrying about invasive species and the ways in which they are changing our world. They are compiling data, making lists, and documenting impacts of invasive species.

What species are on the list? A number of the names will be familiar to most people interested in conservation, ecology, or whole organism biology. There are invasive mammals such as house mice, Carolina gray squirrels, and brushtail possums; birds like the mynah and the English starling; plants including kudzu, purple loosestrife, and prickly pear; insects such as the malaria-carrying *Anopheles* mosquitoes, Asian gypsy moths, and red fire ants; mollusks including zebra mussels, apple snails, and giant African snails; and numerous others—cane toads, Nile perch, brown tree snakes, Dutch elm disease, and the chytrid frog fungus. The list goes on, and it is extraordinary for the breadth of organisms represented and the extent of the geographic distribution of these creatures. But there is one telling and very serious gap in such lists.

The most invasive, most environment-altering species of all—the one that has contributed to or directly caused the extinctions of thousands of species and the alteration of almost every habitat you can think of—is not on the list. If you go to the Global Invasive Species Database (http://www.issg.org /database/welcome/) and search the entire list, not just the 100 worst offenders, and type in "*Homo sapiens*," our species, the following answer comes up: "No invasive species currently recorded."

Why not?

Maybe because we—we humans, *Homo sapiens*—are the ones writing the list and we do not want to face our culpability. As a wise and eponymous possum in Walt Kelly's cartoon, "Pogo," said in a 1971 Earth Day poster, "We have met the enemy and he is us" (see Figure 1.1).[1]

So he is—so we are.

I maintain that humans are the most invasive species that has ever lived. From a modest evolutionary beginning in Africa about 200,000 years ago, our kind has spread across the entire world, invading geographic region after geographic

Figure 1.1. Using the characters from his comic strip *Pogo*, Walt Kelly designed this poster for Earth Day 1971 to show the destruction humans have wreaked on our planet.

region, settling in to exploit new habitats until we can now be found living on every continent. We live in the sweltering tropics, in the far, cold north, on tops of mountains and in deep valleys, on islands and continents and island continents, in deserts, in rain forests, in temperate forests, and in open environments or closed ones. We do not live under water, except in artificial habitats like submarines, but many humans live in boats or on floating villages in lakes and rivers. We have become established in nearly every habitat on earth. It is an awful and awe-inspiring record.

The adaptability, cunning, and technology that make our enormous geographic distribution possible also mean that our invasive success is unparalleled by any other species. We adapt ourselves to an incredibly broad range of habitats, lifestyles, diets, and climates. We adapt by the recording and sharing of knowledge through our exceptional language skills. We do this not only biologically but also through the invention and use of cultural buffers, like clothing, fire, shelter, the capture and transport of water, and the planting of crops. Some of these can be considered tools—items constructed to be used—and others are behaviors. We also create and use tools that have allowed us to circumvent the long process of evolutionary change as a way of expanding our resources and skills. For example, we have not evolved sharp, cutting teeth for slicing up food; we have invented first stone tools, then metal ones, and various implements to facilitate cutting made from almost every substance imaginable.

We have even "invented" and "manufactured" living tools by domesticating other animals, by controlling their breeding until their genes produce the attributes we want. Then— because a domestic species and a domesticating species form a kind of covenant or pact—we are able to borrow some of the domesticated species' anatomical and behavioral skills, such as their superior eyesight and hearing, their speed in locomotion, their tremendous power, their cutting teeth, quick paws, and deadly claws, and their intensely specific sense of smell. This arrangement seems to work to our advantage, but I caution that a nuanced view is more realistic. Living tools like horses, dogs, cats, cattle, or pigs are not passive but active participants. If they do not want to and agree to work with us, they do not work to our advantage at all. Domestication is a continually negotiated agreement between two species, not an enslavement of one by the other.[2] And some species flatly refuse to be domesticated.

Even though the domestication of plants has involved less negotiation, what such plants have given us has been essential to our worldwide success as a species and our ever-increasing population. It has not been a one-way street, however. Plants have benefited from their domestication by increased fertility, larger seeds, protection from animals, or simple boons like watering.

There has been a serious price for our success in worldwide terms. Modern humans, *Homo sapiens*, have decimated millions of acres of once-productive land until the soil has eroded into the seas; this practice continues. We have cut down or burned vast areas of forest and woodland and prairie that once produced oxygen to replenish our atmosphere as well as fruit, leaves, roots, and nuts to feed ourselves and many other creatures. We have single-handedly accounted for the pollution, poisoning, and drying up of innumerable water sources around the world, thanks to our insatiable and growing needs, our toxic chemicals, and our enormous piles of debris. More than all this, we have contributed to the extinction of more species than we can calculate.

I am not alone in laying the blame at our own doorstep. In 2005, ecologists David Burney and Tim Flannery wrote a review article called "Fifty Millennia of Catastrophic Extinctions after Human Contact."[3] The title tells most of the story. Time after time, in place after place, humans have spread into new areas, and shortly after our appearance, numerous other species have gone extinct. As Burney and Flannery maintain, there is "a global pattern of human arrival followed by faunal collapse and other ecological changes . . . without known exception."[4] Think about that: a *global pattern* of faunal collapse. In many cases, the now-extinct parts of the fauna included many large mammals, birds, or reptiles. Large animals are more vulnerable to extinction because they reproduce more slowly than small animals, thus making the loss of any reproductive-age adults more serious. And larger animals need larger ranges or territories to live in, so they are more vulnerable to habitat loss, too. Since Paul Martin started writing about this pattern at length in 1967, the phenomenon has been called "megafaunal extinction."[5] The devastation we have wrought does not stop with animals but includes plants, too. In another recent article, an impressive team of ecologists writes: "Modern extinctions are largely being caused by a single species, *Homo sapiens*, and . . . from its onset in the late Pleistocene, th[is] . . . mass extinction has been characterized by the loss of larger-bodied animals in general and of apex consumers in particular. . . . The loss of apex consumers is arguably humankind's most pervasive influence on the natural world."[6] An apex consumer is one at the top of the heap, a species that directly or indirectly consumes all the nonpredatory biomass in the ecosystem. We are apex consumers, of course, and we have deliberately or unconsciously worked hard to eliminate any other apex species that might compete with us in every ecosystem we have entered.

In some historical instances, climate change disrupted ecosystems and initiated an apparent population decline in some large mammals—leaving them especially vulnerable to human activities—yet there can be little doubt that human invasion was a (but not necessarily the only) key factor in triggering extinctions on the large landmasses of Eurasia, Australia, and the Americas, in the circumpolar regions, on numerous islands, and on the island continent of Madagascar.[7] I would not claim that the arrival of humans has been the only factor contributing to many extinctions, but a spate of extinctions following human settlement is the norm.

Apex predators are especially influential and powerful in shaping an ecosystem. Marine ecologist Robert Steneck of the University of Maine observes, "Studies from widely divergent ecosystems all found a single predator can control the distribution, abundance, body size and species diversity in the system."[8] The human role in extinctions is easiest to prove on small islands where the limits of resources vital to the survival of the indigenous species are more obvious. But even on a continental basis, there is good evidence of the human role in the loss of mammoths and mastodons, of aurochs and wild horses, of many larger Madagascan lemurs, of dodos in Mauritius, and of moas in New Zealand. We humans have helped kill off amazing giant marsupials, enormous predatory birds and lovely passenger pigeons, giant ground sloths, woolly rhinos, fearsome dire wolves and saber-toothed cats, as well as other species that managed to survive changing climates—until humans came along. Human predation is not the sole factor in all of these extinctions, but we cannot escape the conclusion that we have played a crucial role in them.

This book looks at a particularly pivotal point in our history, the time during which the last species of nonhuman hominin (Neanderthals) went extinct. I argue here that Neanderthal extinction was triggered by the appearance of modern humans in Neanderthals' geographic range; simply put, humans are a supremely well-adapted invasive species, and we behaved exactly like one during this extinction event. I have been pleased to find other anthropologists expressing similar views, and I try to give full credit here to others when credit is due.[9]

Why our close cousins the Neanderthals died out has puzzled paleoanthropologists ever since Neanderthals were first recognized as a species in 1856. As more fossils, modern research, and new techniques of analysis have proliferated, the extinction of a species with such familiar attributes—with the ability to make fire, to manufacture and use tools, to cooperate socially, to take down large mammals, to use symbols, art, and communication to at least some extent—has continued to seem enigmatic. But once you understand what makes a successful invasive species and which factors determine its impact on the ecosystem it invades, you will puzzle no longer.

2

HERE WE COME, READY OR NOT

What exactly is an invasive species? Perhaps the simplest definition is that it is a species that moves into a new geographic region where it has not previously (historically) lived. The species that are invaded—those that are indigenous to an area without human intervention—are considered native. An endemic species is a subset of native ones, comprising those that have evolved in a particular area and are found only there. An invader is neither of these; it is a species that does not "belong" in the region. An invader is nonindigenous, nonnative, nonendemic, alien, and very often disruptive to the ecosystem.

The U.S. government legally defines an alien species as being "with respect to a particular ecosystem, any species, including its seeds, eggs, spores, or other biological material capable of propagating that species, that is not native to that ecosystem."[1] By this same executive order, an invasive species is defined as "an alien species whose introduction does or is likely to cause economic or environmental harm or harm to human health."[2] What this means is that a species that is not disruptive to the ecosystem or harmful to humans is not technically considered invasive. This is a somewhat shortsighted and anthropocentric point of view.

How is a species' invasion different from a geographic expansion? The distinction lies in a question of timescale, distance, and influence.[3] Expanding a species' territory by a few miles probably will have a negligible impact on the ecosystem as a whole. Expanding a species' distribution from one habitat or ecosystem or continent to another—across a mountain range or other formidable geographic barrier—is considered an invasion, as this change is more likely to have far-reaching effects.

Ecosystems are complex entities, crisscrossed and bound together by webs of cooperation, symbiosis, and mutual dependencies. The close relations among species within an ecosystem are stretched, warped, and complicated by competition, mutually exclusive needs, complementary needs, and the uncertainties of coexistence. Dropping a wholly new organism into the mix can disrupt the functioning of the entire system, unless the new organism is so ill adapted it dies out without becoming established as a viable, continuing population.

An invasive or nonindigenous species is easily defined in theory but not in practice. If a native or indigenous species is one that has been part of the local ecosystem for a considerable length of time, the key question is: How long? How many years, centuries, or millennia must a species be part of an ecosystem to be considered native?

In the United States, 500 years in a particular ecosystem is often used as the Rubicon that divides native from nonnative, alien, or invasive. Ecologists who focus on living organisms tend to think in a very short time frame, perhaps a few human lifetimes, and 500 years is a convenient dividing line. I regard this short time frame as a weakness in a very strong field. Since its inception, invasion biology has been closely tied to conservation efforts and strategies, so changes that occur within a relatively short time span garner lots of attention. Such changes can be seen, measured, and documented. It is much harder to see and understand the causes of an invasion when your time frame is thousands or tens of thousands of years.

In contrast, paleontologists and paleoanthropologists, like me, are far more comfortable thinking in terms of hundreds of thousands of years. The advantage of considering a long time span is that great evolutionary and biological truths and trends can be seen clearly. Are animals that evolved in Asia and entered the Americas—such as humans, mammoths, horses, and wolves—considered indigenous if they have been on this continent for 20,000 years? What about 10,000 years? A mere 500 years is not long enough to see massive effects from an evolutionary perspective except in very short-lived species that have short generation times. The greater time depth that is involved in evolutionary studies means that I think about invasive species and resulting extinctions slightly differently from the way most invasion biologists do. The weakness here is that many short-term fluctuations—changes that may last hundreds of years and wreak havoc on individual species—can be very hard to detect in geological time. We paleontologists and evolutionary biologists may take too long a view, in some instances. On the whole, I believe understanding the principles of invasion biology will help resolve some long-standing questions in evolutionary biology. Sometimes looking at a short-term effect provides the key to understanding a long-term one.

To properly explain the hypotheses and conclusions that underlie this book, I have to be perfectly clear what I mean by an invasive species. The working definition I use goes beyond moving into a historically new territory. One criterion for determining whether a species is invasive or not often involves its impact. Invasiveness is not an all-or-nothing event. To qualify as an invasive species, an organism must progress through a series of sequential steps.

First comes arrival: the alien species must get to a new ecosystem by some mechanism, which in modern times often involves humans. Invasive species hitchhike, catching rides on boats or planes or cars, lying hidden in bilge water, in containers, in personal luggage, in the folds of our clothing, or inside our bodies. Sometimes larger invasive species are brought along purposefully by humans who view them as a kind of insurance that they will have in new habitats the species of animals that are important to them. As an example, people have carried not only domesticated animals but also deer, moose, rabbits, and innumerable plants around the world. Inadvertently, we have also spread many insects and parasites. We humans do not like to travel alone.

Second, the invader must establish itself as a viable wild population. As a ballpark figure, the smallest population of mammals—known as the minimum viable population (MVP)—is usually considered to be 1,000 individuals. If there are fewer than that, the lack of genetic diversity may lead to a high frequency of deleterious mutations because of inbreeding. Small populations are also more vulnerable than large ones to random events like hurricanes, disease epidemics, or drought. A small founder population can be wiped out in a very short time by such events. But an article published in 2007 on this topic suggested that an MVP lower than 1,000 individuals might be adequate.[4] My concern is that an MVP is routinely defined as one that has a 95 percent chance of being around 100 years later. One hundred years? That is an undetectable blip in paleontology, not long-term survival, though 100 years is a long time in human experience. (A human generation is generally calculated as twenty years, so 100 years is five generations. Few of us can trace our family tree so far back.) From my perspective, if you are trying to calculate long-term survival, 1,000 individuals may be way too few because random chance, local or global catastrophe, or minor change can eliminate a 1,000-individual population easily.

Let's consider a good analogy for the case of modern humans invading Eurasia: the peopling of Australia. What is advantageous about doing this is that there were absolutely no humans or hominins of any kind on the Australian continent prior to the arrival of modern humans, so there is no question of confusing archaeological sites or tools made by one species of hominin with those made by another. Such confusions are often a problem in Eurasia. Also, the peopling of Australia occurred in a time frame of 48,000–46,000 years ago, just about when modern humans first got to Eurasia as well. We can expect that the first modern humans to get to Australia had abilities roughly comparable to those of the first modern humans to reach Eurasia,[5] with the added proviso that the first human invaders to reach Australia must have had watercraft and considerable knowledge of seafaring.[6]

So what happened? And how many were needed to people Australia?

A recent study by Alan Williams of the Australian National University calculated the number of humans who invaded Australia about 45,000 years ago in a clever way.[7] It has usually been assumed by anthropologists that a small number of people, perhaps 50 or 100, colonized Australia. Williams set out to test this assumption mathematically. First he compiled a database of every radiocarbon-dated, well-documented archaeological site in Australia prior to European contact, which amounted to just over 5,000 sites. Then he used the numbers of sites through time as a proxy for population numbers, based on the reasonable premise that more people leave more sites. This allowed him to calculate relative growth rates of the human population over time, even though he did not know how big the initial or founder population was. He could ask how long it took for the founder population to double, or triple, or increase tenfold. Once he knew the population's growth rate, he could calculate backward from the size of the Aboriginal population at European contact (in about 1788) to see how big the founder population must have been. Estimates vary between 770,000 and 1.2 million Aboriginal individuals in 1788. The number is not precise because European contact brought European diseases, which decimated many Aboriginal groups within just a few years—not to mention the deliberate killing of many Aborigines by European settlers. And no systematic census of the entire continent was conducted either. Still, working from good estimates of population densities in various areas, the early European settlers got a reasonable idea of the numbers of Aborigines in Australia.

Surprisingly, Williams found that if the founding population of Australia had been between 50 and 150 individuals—as had always been assumed—then the Aboriginal population would have been no larger than 19,000 at European arrival. Although early population estimates of Aborigines in Australia may have been imprecise, they would have had to have been insanely inaccurate to mistake fewer than 20,000 individuals for more than half a million. Clearly, a founding population of 50 to 150 Aborigines was not adequate to produce the observed population of 770,000 to 1.2 million Aborigines at European contact. In fact, to produce the Aboriginal population numbers at contact that fit the historical records of the contact period, the founding population must have been much bigger, between 1,000 and 3,000 individuals.

This large number of founders bespeaks intentionality, according to Williams, and impressive watercraft. "It's not just a family that got stuck on a raft and washed away," he says. "It's people with the intention to move, to explore."[8] Since Australia was separated from the next nearest landmass by at least 80 kilometers (km) of open ocean, watercraft must have been used to transport the immigrants. Prior to invading Australia, Aboriginals probably relied on seafaring for obtaining food on a regular basis. Because the number of Aborigines reaching Australia is so much larger than had been assumed—and yet not much larger than an ecologist would predict for an MVP—the Aboriginal boats observed by Europeans at contact seemed too small and not sufficiently sturdy to have carried such a large number of people such a distance. Had Aborigines forgotten some of their seafaring skills in the thousands of years between their invasion and that of modern Europeans? Apparently. The record suggests that the fishing and seafaring habits of the founding Aborigines were largely abandoned for a greater reliance on terrestrial resources once the original Australians reached the continent. Perhaps the skills in boat building were not needed in the face of the rich Australian fauna that was naive to the deadly ways of humans.

Clearly not all invasive species intend to invade, nor are they always self-transporting, but humans have been on at least a few occasions. Getting there is far from the only challenge for an invasive species, although (as in the case of Australia) the reality of invading a new territory, whether deliberately or accidentally, can be difficult. To be a successful invasive species, the new population must spread well beyond its point of introduction.[9] This is where Williams's extensive database of Australian sites confirms a key point. The early archaeological sites in Australia are spread over far-flung areas of the continent, showing that humans moved from their point of entry, spreading into suitable habitats everywhere on the new continent, probably by following coastlines and rivers. This spread implies population growth, as 3,000 people can't very well inhabit the entire continent. Humans entering Eurasia appear also to have spread across their new continent very rapidly, in geological terms if not in human ones.

Not all nonindigenous species complete all of these steps of landing, surviving, spreading, and increasing in population size to become viable in the long term. In fact, a commonly cited idea in invasion biology is the "rule of tens." Simply put, this concept states that only 10 percent of all species expand beyond their native range—you can think of these as "traveling species"—through their own actions or those of other species. Of these newcomers, only about 10 percent will move into the wild. These are traveling and expanding species. Only about 10 percent of these expanding species will establish themselves as viable, wild populations. Finally, only about 10 percent of those will become pest species that disrupt the ecosystem or, in conservation terms, threaten or compete with indigenous species.

So, for example, if you start with 1,000,000 species, only 100,000 will move into a new range, only 10,000 will spread into the wild, only 1,000 will become established as a viable population, and only 100 will become pests. Thus, we make a lot of fuss about very few species that undergo a very difficult process.

Why? Because the record of our planet shows that a very few species can trigger highly significant changes. Many ecologists consider invasive species to be one of the top five drivers of extinctions. Others are climate change or habitat disturbance, pollution, disease, and overexploitation (by humans). In reality, these factors are interactive and synergistic; rarely if ever does one of these influences operate in isolation to cause an extinction. In fact, once one of these factors is at work, an ecosystem may become more susceptible to other dangerous factors, particularly invasive species. The interactions and complex vulnerabilities make it difficult to parse the relative role of invasive species as a potential cause of extinction.

There have been five great global extinctions caused by massive events like volcanic eruptions or the asteroid impacts that triggered the end of the dinosaurs. Some would say that humans are causing the sixth great extinction even now.[10]

But if you ask whether we have evidence that invasive species cause one-by-one, everyday extinctions, the answer is "yes." For example, in an influential paper published with colleagues in 1998, David Wilcove, a senior ecologist at the Environmental Defense Fund in Washington, DC, considered a total of 1,880 U.S. species that were judged by the Nature Conservancy, the U.S. Fish and Wildlife Service (USFWS), or the National Marine Fisheries Service to be imperiled, threatened, or endangered. Gathering specific information about each species or population let Wilcove's group rank the causes of threatened extinction across all species.[11] The most ubiquitous was habitat loss or destruction, which affected 85 percent of the species. It's pretty simple. A species cannot survive if it does not have enough land to live on. Forty-nine percent of the species were threatened by alien or invasive species that posed competitive threats by using the same resources. Twenty-four percent of the species were threatened by pollution in their habitats, which is really just a particular type of habitat destruction. Seventeen percent were threatened by overexploitation by humans, something we can document in the populations of Atlantic cod, for example. Only 3 percent of the species in this study were threatened by disease.

Of course, the primary culprits in habitat destruction in the last few centuries are humans, whose extractive industries (such as logging, fishing, or mining), agriculture and livestock ranching, and development of infrastructure, including roads, buildings, pipelines, dams, and reservoirs, damage or destroy habitat. Thus, many of the causes underlying habitat loss or degradation can be attributed to a single species: ourselves, modern humans.

Another analysis of extinction causes was carried out by Miguel Clavero and Emili García-Berthou, both of the University of Girona.[12] They compiled statistics on all 680 extinct or nearly extinct animal species that are on the IUCN Red List. Of these, 170 (25 percent) are associated with a particular extinction cause. One of the commonest extinction causes cited by the IUCN for 54 percent of the species (91 cases) is an invasive species. The results of this and other studies lead to the same general conclusion: invasive species are a major contributing factor in many and possibly most extinctions. In the long paleontological perspective, we humans must be considered invasive in any locale except Africa.

Recognizing *Homo sapiens* as an invasive species helps explain a lot about our past and current evolutionary position. Invasion biology offers us new tools with which to examine our evolutionary history and illuminate our place in nature.

Invasion biology has blossomed as a field of study since 1985, when ecologist Charles Elton of Oxford University published his landmark work, *The Ecology of Invasions by Animals and Plants*.[13] This book brought to light a fascinating paradox: the appearance of a new species in an ecosystem can have either far-reaching effects or very little effect at all. In the search to understand how invasions function and which outcomes occur, researchers have studied everything from sessile marine organisms to insects, plants, fish, birds, and mammals and their response to invasion.

In the process, terminology has been refined, principles have been articulated and tested, and the tone of the writing has shifted. Elton wrote of "alien species," "foreigners," and "invaders"; now ecologists often write more neutrally of "colonizers," "species introductions," or "nonnatives" instead. Though I understand that emotion-laden terms may hinder the objective evaluation of information, I choose to use them here because such terms underscore the harsh reality of the changes that can be wrought by a single new species. Using bland, neutral terms runs the risk of obscuring the true impacts of invasions.

Nobody expects disparate creatures with different habits, diets, reproductive, and locomotor abilities to behave exactly the same under situations of invasion. A new nematode may not affect an ecosystem as dramatically and widely in a short period of time as, say, a new and deadly mammalian predator. However, some fundamental principles by which invasions generally proceed are becoming clear and can be useful to us in examining our survival as a species.

With caution, attention, and intelligence, we can use the methods of invasion biology to look at ourselves and our evolutionary history. This relatively new field offers us a new perspective on some of the greatest questions in human evolutionary studies: Why did our particular lineage survive as *Homo sapiens*? Why and how did the last widespread "other" hominin species, *Homo neanderthalensis*, go extinct in the very same habitat in which the newly arrived *Homo sapiens* thrived?

Not all paleoanthropologists would feel comfortable designating Neanderthals as *Homo neanderthalensis* (a separate species), as I do in this book, rather than *Homo sapiens neanderthalensis* (a subspecies of modern humans). My point is not to engage in taxonomic arguments but to make clear to the reader which group I am writing about at any given time. There are certainly morphological, or shape, differences between Neanderthals and modern humans in many parts of the body. Anyone can readily be taught to recognize modern human skulls from those of Neanderthals; those people with special interests can usually attribute additional parts of the skeleton to different species with little difficulty. There were also cultural differences between Neanderthals and modern humans. So whether the two were separate species or simply highly distinctive subspecies is not the most important issue for my discussion here.

Another potential source of confusion lies in spelling and pronunciation. The common name for Neanderthals is just that, or possibly Neandertals. The former, which I use here, is derived from the species name given to the fossils in 1861, *Homo sapiens neanderthalensis*. Later, German spelling was revised to reflect pronunciation, which is properly "neander-TAL." The Latin name of the species cannot be altered to correlate with the new spelling, but some anthropologists use Neandertal nonetheless. The important point is to know the sort of creature about which you are speaking or writing.

Genetic evidence from Neanderthal bones has revealed some of the thorny problems in classifying two groups that are closely related. A species is generally defined as a biological population within which there is interbreeding but which is separated reproductively from other similar groups. This means that a member of one species does not hybridize and produce fertile offspring with a member of another species. In practice, this definition can be very difficult to apply.

Cliff Jolly, who was my adviser for my doctoral thesis, has spent much of his career studying hybrid zones where two of the five generally accepted species of baboons overlap—and hybridize. If over time the hybrid zones and the extent of hybridization remain roughly constant, then the two species of baboons may indeed be good (valid) species. Alternatively, if hybridization is occurring more commonly and not negatively affecting survival, then the two putative species may be in the process of separating from each other, of forming truly separate species.

Hybrid zones are not particularly rare and are known to exist between species of toads, salamanders, birds, and many mammals other than baboons. In fact, there are many cases where humans interfere and deliberately encourage breeding between captive animals that either would not meet or certainly would not interbreed in the wild. What are we to make of lygers, part lion and part tiger? There is no evidence of them in the wild, so perhaps this is a moot point. But biology is a messy and complex science by which species form, change, and die out constantly, if only you have a long enough time frame.

Because the genome of each lineage within a species mutates and changes throughout time, drawing hard and fast lines between species is tricky. Many primate genera have more than one species in them; for example, the South American monkey genus *Callicebus* includes twenty-nine separate species. But the state of classification used today is partly a function of which group has survived. Survival of a species is in turn affected by its adaptability, the specificity of its habitat and food preferences, the loss or creation of habitats needed by the species, climate change, interaction with other species, and many complex factors. Humans are peculiar in having only a single surviving genus and species alive today, whereas gorillas, chimpanzees, orangutans, and gibbons—the apes—each have more species. But other primates are also monospecific.

It was hoped—fervently by some—that advances in genetic studies would resolve the problem of species identification for us. The first set of problems revolved around the sheer difficulty of extracting undistorted DNA from an ancient bone. The first attempts involved mitochondrial DNA (mtDNA), which is present in each cell in the body in 100 to 1,000 copies. These large numbers increase the chances that stretches of intact mtDNA can be recovered and then stitched together to give a complete mitochondrial genome. The possibility of contamination with the mtDNA of modern humans—from the excavators, from the geneticists themselves, from technicians, curators, and paleoanthropologists who handle ancient bones—was another huge problem.

Eventually, with great care, by splitting samples and having two labs conduct analyses independently, a large part (400 base pairs) of an mtDNA genome from a Neanderthal was recovered in 1999. It was a real breakthrough. As geneticists developed better and better techniques, more genomes were extracted and published. With the first group of mtDNA genomes—up to about ten individuals—there was no overlap whatsoever between Neanderthals and modern human genomes. As geneticist Matthias Krings of the Max Planck Institute of Evolutionary Anthropology wrote in the first paper published on Neanderthal genetics: "These results indicate that the Neandertal mtDNA gene pool evolved for a substantial time period as an entity distinct from modern humans and give no indication that Neandertals shared mtDNA with modern humans."[14] In other words, there was no mistaking the Neanderthal mitochondrial genome for a human one; they were entirely separate and different, according to the early results.

Mitochondrial DNA also offers the advantages of being shorter than nuclear DNA, and it is passed along only the female line. Whatever mtDNA you, I, or anyone else possesses, it came from our mothers. Our fathers made little or no contribution to my or your mtDNA. That is because the sex cell derived from your mother, the ovum, has both a nucleus and a cytoplasm. The cytoplasm is where the organelles known as mitochondria, which provide energy to the cell, reside. They also house the mtDNA. The male contribution to the fetus is a sperm, which is predominantly a nucleus with a tail attached. There is little chance for mtDNA to be passed from father to offspring because the father's sperm has very little cytoplasm and hence few mitochondria.

With improved techniques, a complete draft sequence of Neanderthal mtDNA was published in 2008.[15] The authors announced that this work "unequivocally" established that the Neanderthal mtDNA falls outside the variation seen in living human mtDNA. Of course, because of the mechanism by which mtDNA is transferred from parents to offspring, an mtDNA lineage is carried only by a mother, passed on to her daughter, passed on again to her daughter's daughter, and so on. The male offspring of the original mother will have the same mtDNA as she did but no ability to pass it to their offspring. If all of the daughters of the original mother fail to have daughters, the lineage will be lost; it will become extinct. This is the fate of the large majority of mtDNA lineages, so saying the mtDNA lineages in Neanderthals are not found in modern humans does not mean there was no interbreeding. It simply means that, if there was interbreeding, the Neanderthal females' mtDNA did not survive.

The estimated time of divergence between the Neanderthal mtDNA and the modern human one lies between 646,000 and 800,000 years ago. Estimating the divergence time between two genomes sounds simple but isn't. The basic idea is that base pairs of amino acids that make up mtDNA (or nuclear DNA) mutate at random at a steady rate. If you know the mutation rate and then count how many differences occur between the genomes of two species, you can easily calculate when the two species diverged from each other genetically. But not all genes mutate at the same rate or even at a steady rate over time. Species with short generation lengths mutate faster than those with long generation times, for example. Particular genes are more prone to mutation than others, too. This means that the more genes you can compare between two species, the more likely you are to estimate the divergence time correctly. But it is an imprecise form of dating because nothing is actually being dated. What estimating divergence times does is calculate how much time probably elapsed to create the degree of genetic distance between two forms.

In time, various labs began attempting to sequence the nuclear DNA of Neanderthals, based on the chromosomes—one from the mother, one from the father—inside the nucleus of each cell. Problems in dating divergence times from nuclear DNA aside, other difficulties arose. Until geneticists were able to demonstrate that they could retrieve nuclear DNA from mammal bones from the same sites as Neanderthal bones, many curators were reluctant to part with pieces of irreplaceable Neanderthal remains.

Top labs like the one in the Max Planck Institute in Leipzig, led by Svante Pääbo, had developed reliable and exacting methods for extracting mtDNA from ancient bones. To help persuade curators to let them sample more specimens, without undue risk, they turned to the bones of other mammals, starting with cave bears from the Croatian site of Vindija. Time after time, they failed to extract anything from the cave bears. Pääbo's team then turned to trying to extract nuclear DNA from mammoths recovered from the permafrost, hoping that such preservation might be less destructive of DNA. It was, and within a few years, various technical breakthroughs and an influx of very generous funding meant sequencing of ancient nuclear DNA could be carried out more rapidly and with fewer tedious procedures. Nuclear genomes are the ones ordinarily referred to when scientists speak of DNA; one chromosome in the pair comes from the mother and carries her DNA, and the other is contributed by the father and carries his. Thus in each generation, there is a recombination of genes in the new individual.

As Pääbo's group sequenced more and more of the Neanderthal nuclear DNA, a surprising fact emerged. Though Neanderthal mtDNA did not overlap with modern human mtDNA, nuclear DNA from Neanderthals did overlap with that known from modern humans. But the overlap was small—perhaps 1 to 4 percent—and appears to be confined largely to modern humans of European or East Asian ancestry. Neanderthal genes are more rare or absent in humans of African ancestry. This fact has suggested to some researchers that the interbreeding occurred after early modern humans left Africa, when they encountered Neanderthals in the Levant or in Eurasia, but the exact meaning of this shared genetic information is contested.

Richard Green is part of the team at the Max Planck Institute of Evolutionary Anthropology. Initially, Green told a reporter from *Science News* that the extent of overlap was so small that nothing "genetically important" may have been transmitted by interbreeding. He said, "The signal is sparsely distributed across the genome, just a bread crumb's clue of what happened in the past. . . . If there was something that conferred a fitness advantage, we probably would have found it already by comparing human genomes."[16] Another team member, Montgomery Slatkin of the University of California, Berkeley, remarked, "We don't know if interbreeding took place once, where a group of Neanderthals got mixed in with modern humans, and it didn't happen again, or whether groups lived side by side, and there was interbreeding over a prolonged period."[17] Many questions still remained unresolved about Neanderthal and modern human genetics. More surprises were in store.

Theoretically, separate species are not able to crossbreed and yield fertile offspring. Thus, on the face of it, the evidence of crossbreeding between Neanderthals and modern humans argues against a species-level separation. However, the highest degree of inbreeding anyone is willing to postulate is between 1 and 4 percent. Such a low level of crossbreeding may have had little substantial biological effect on either group and is paralleled, for example, by crossbreeding among modern species of baboons. Also, the total number of Neanderthal genomes that have been investigated is terribly small, and the number of modern humans that have been genotyped is also small relative to a world population of modern humans that is over seven billion individuals.

Anders Eriksson and Andrea Manica of the University of Cambridge have looked at the Neanderthal DNA from another perspective.[18] They argue that the overlap between genomes of modern humans and Neanderthals may come from the ancient common ancestral population that gave rise to both species, not from interbreeding after the two had diverged genetically. So far, there is no firm basis for choosing one interpretation over the other.

Two recent papers have explored the overlapping sequences to analyze where and how much Neanderthals and modern humans differ in their genomes. By looking at 1,004 genomes from modern humans, a team from the Max Planck Institute

was able to demonstrate that inclusions from the Neanderthal genome were not randomly scattered throughout the
modern genome but occurred preferentially in certain areas
of the DNA.[19] Those areas are rich in genes that affect keratin, the protein involved in skin, nails, and hair. This suggests that these genes may have helped modern humans adapt
to the colder Eurasian climate even though Neanderthals
themselves preferred warmer to colder areas. Another possible interpretation is that the Neanderthal genes associated
with keratin had to do with wound healing,[20] as keratin helps
block pathogens.

Not all Neanderthal genes seem to have been useful to
modern humans. In fact, several Neanderthal alleles identified by the Max Planck team are linked to diseases, medical
conditions, or behaviors linked to medical problems, including lupus, biliary cirrhosis, Crohn's disease, predilection to
smoking, and type 2 diabetes. Still more interesting was the
fact that long stretches on modern human genomes have no
evidence of Neanderthal genes, suggesting that the genes
coded for in these sections were actively selected against because they were deleterious to modern humans. The team refers to these sections as Neanderthal ancestry "deserts." The
largest "desert" is found on the X or female sex chromosome,
where genes associated with lowered male fertility are often
found. Confirmation of this conclusion was offered by investigating the frequency of Neanderthal alleles in genes that
are tissue specific. Genes affecting the testes, an organ that
is directly involved in male fertility, were found to have strikingly few Neanderthal alleles. The implication is that male
Neanderthal–human hybrids were either infertile or had diminished fertility. The two species were very nearly incom-

patible genetically. No wonder the extent of cross-breeding between the two species appears to have been so low.

Almost simultaneously, two researchers from the University of Washington, Benjamin Vernot and Joshua M. Akey, published the results of their study.[21] Looking for Neanderthal genes, Vernot and Akey inspected the genomes of 379 modern Europeans and 286 East Asians. They, too, found genes involving keratin and skin pigmentation among others. Although the function of all the Neanderthal genes is not yet known, perhaps the most astonishing result of their investigation was the finding that as much as 20 percent of the Neanderthal genome still persisted in some humans. This finding does not mean that the extent of interbreeding between Neanderthals and modern humans was much higher than had been estimated but simply that subsequent to interbreeding, some Neanderthal genes were so deleterious as to be eliminated by the forces of natural selection while others were less harmful and persisted.

All of this information does not help us decide how to classify Neanderthals and modern humans taxonomically. The genomes tell us only that Neanderthals were close enough to exchange genes with modern humans to a limited extent. The biological similarities between Neanderthals and early modern humans were striking but not a perfect match. Culturally, many of the behaviors we humans like to think of as special or advanced were present in Neanderthals, too. But behaviors are not simply a matter of genes but a complex and often idiosyncratic mixture of genetics, learning, and luck. There is no gene for making a spearpoint, for example, or for catching a mammoth. By the time our early ancestors encountered Neanderthals in Eurasia, Neanderthals were clever, adept, and well adapted to their environment and ecosystem—and yet they went extinct and we did not.

Why? Why did we survive while they went extinct? Why are we now alone and yet nearly ubiquitous in the world? And what does this reveal about our relationship to Neanderthals and our invasive impact on the world?

3

TIME IS OF THE ESSENCE

When I started writing this book, paleoanthropologists accepted that both modern humans and our close relatives, the Neanderthals, lived in Eurasia between about 50,000 and 25,000 years ago. Archaeological sites attributed to Neanderthals and those made by modern humans are often close together; sometimes the two groups seem to have alternated use of a particularly attractive cave or rock shelter. Modern humans and Neanderthals were much alike in many ways: they were large-bodied, intelligent, adept hunters of large game, tool makers, and group dwellers. Both had fire. They both may have had language (though this is an enormous debate in itself), and they both created enduring signs and symbols that can be called communication or artwork, although humans did this much more often than Neanderthals.

What clearly differentiates the two is morphology—that is, anatomical shape—and bodily proportions. Their skulls are simple to distinguish if they are mostly complete, with Neanderthals having bigger brow ridges, more elongated skulls, a particular bony protuberance on the back of the skull known as a Neanderthal "bun," and a face that from nose to upper jaw is thrust forward. Accompanying this is a more forward placement of the teeth, which differ statistically in size and in some shape details from those of modern humans. The Neanderthal middle face also is built to anchor a larger, more prominent nose and has a larger nasal opening. Bodily, Neanderthals were bulkier and much more muscular, with larger joints and thicker bones showing strong muscle markings. Neanderthals weighed in at about seventy-eight kilograms (kg) for males versus sixty-nine kilograms for early modern human males; female Neanderthals were sixty-six kilograms versus an average of fifty-nine kilograms for female modern humans. These bodily differences reflected differences in hunting technique, daily exertion, and probable means of heat retention.[1]

A key concept in an endeavor like this is realizing that different evidence has different importance. Where there are good skeletal remains, telling a Neanderthal from a modern human is not a difficult task. When remains are poorly preserved, damaged, or fragmentary, the task is more difficult. Where there are no hominin remains, paleoanthropologists often rely on the premise that modern humans made Upper Paleolithic stone tools (classified into discrete and often sequential industries with names like Aurignacian, Gravettian, or Magdalenian) and Neanderthals made different constellations of tools, often favoring different techniques. Generically, these groupings are called tool industries, and those typical of Neanderthals are known as the Mousterian industry or the earlier Mousterian of Acheulian tradition. These industries are distinguished by the detailed techniques used to make the tools, the proportions of different types of tools that were presumably made for different purposes, and sometimes by the presence of a particular distinc-

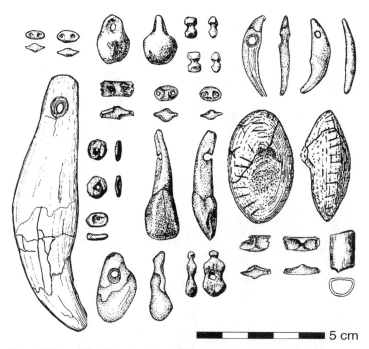

Figure 3.1. Upper Paleolithic sites often yield jewelry or ornaments made of animal teeth, pierced or scored to be hung from a cord or thong. These specimens come from Aurignacian sites in the Swabian Jura valley of Germany.

tive tool. The presence of tools made out of bone or ivory and of objects of personal adornment—pierced animal teeth or shells, for example, that seem to have been strung on thongs or fibers—are usually taken as diagnostic of the Aurignacian or some other Upper Paleolithic, modern human industry (see Figure 3.1). So are works of art, like drawings, sculpture and paintings, and musical instruments.

These classifications of stone tools into industries began in the nineteenth century, and inevitably, their recognition and analysis have become more complex with time. A key issue

Figure 3.2. These remains from the Grotte du Renne site in France epitomize the controversy over the identification of the makers of the Châtelperronian. If they were made by Neanderthals—the traditional interpretation—then these objects indicate the use of skills and symbolism very like those used in Upper Paleolithic sites (compare with Figure 3.1).

is the Châtelperronian industry, which falls chronologically between the Mousterian and the Aurignacian in some parts of Eurasia. While Mousterian tools consistently co-occur with Neanderthal remains (when there are any skeletal remains at

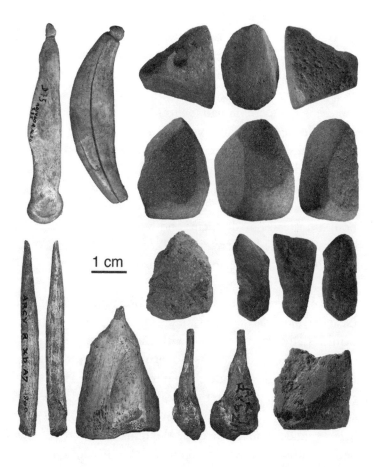

1 cm

all) and the various Upper Paleolithic industries co-occur with modern human remains, the Châtelperronian is less clearly linked to either species.[2] Châtelperronian sites also contain some objects of personal adornment (jewelry) and bone or ivory tools (see Figure 3.2).[3] Are those features alone sufficient to identify the makers of the Châtelperronian as modern humans? Or could Neanderthals have created such objects, too? We simply do not know.

Mixing of remains from different time periods has been demonstrated at some key sites, and the associations between skeletal remains and artifacts is not clear. Several prominent archaeologists have suggested that the Châtelperronian might represent a modern human tool tradition; others suggest it is a Neanderthal tool tradition. The argument is far from settled. Because there are many more stone tool sites than those that include hominin remains, assigning a site to humans or Neanderthals greatly increases the data available for analysis. But there is a catch. Tools do not carry labels indicating which hominin species made them, and hominins may well have borrowed or tried to copy the tools of another species. What then? The most reasonable answer is to be very careful and skeptical of our assumptions.

Since the overlap in time and space of modern humans and Neanderthals was first noticed, paleoanthropologists and lay thinkers alike have wrestled with the interpretation of the facts. Did the expansion of modern humans' territory into Eurasia force the Neanderthals into extinction? How could this have happened, given that Neanderthals had already lived successfully in the Eurasia region for at least 200,000 years before modern humans arrived? Wouldn't Neanderthals have had the advantage over newcomers who did not know the landscape or the fauna?

If humans pushed Neanderthals into extinction, we should be able to find evidence of the process and identify the human advantage. If not, what other factor was at work causing Neanderthals to die out after hundreds of thousands of years? An overlap of 25,000 years seems like a ridiculously long time if the competition between the two types of hominin was acute, especially compared with events that have been studied by modern invasion biologists. From the paleontological perspective, this hypothetical period of overlap may have been much more rapid depending on how quickly modern humans spread across Eurasia and how large the invading and indigenous populations were. The actual overlap between the two species may have been very short, indeed. Time is of the essence.

The basic chronological framework of this problem shook like an earthquake zone early in 2013, when the first results of a massive program to redate many paleoanthropological sites in Eurasia were released. Led by Rachel Wood of the Australian National University and Thomas Higham of the Oxford University Research Laboratory for Archaeology and the History of Art, the team developed highly refined techniques to redate samples from eleven archeological sites in the Iberian Peninsula.[4] The special importance of these sites was that they formed the basis of a claim that, after the arrival of modern humans and a coincident downturn in climatic conditions, Neanderthals abandoned much of a broader Eurasian territory that they had occupied for millennia and retreated to refugia along the Mediterranean with milder climates. In Iberia, they lasted longer than elsewhere, until about 25,000–26,000 years ago, according to this theory and earlier dating techniques. Scholars such as Clive Finlayson of the Gibraltar Museum and João Zilhão of the University of Barcelona have emphasized the role that changing climate and environment might have had on Neanderthal extinction and human survival. Whatever explanation is offered, having a correct and accurate chronology in the more northerly areas where modern humans and Neanderthals overlapped as well as in the later, milder Mediterranean areas is crucial.

How are sites dated? A sample—bone, charcoal, wood, or shell—is taken from the site for dating. If the sample in question is likely to be 50,000 years old or less, radiocarbon dating is appropriate. The underlying premise of radiocarbon dating is that the radioactive isotope of carbon (^{14}C), which is constantly being made in the upper atmosphere, degrades over time into a different isotope, ^{12}C, so the proportions of the two isotopes in an organic sample reveal when that organism was last alive and incorporating carbon from the atmosphere into its tissue. If the sample is older than about 50,000 years, then there is not likely to be enough ^{14}C left to be measured accurately, and other forms of radiometric dating are used. This situation rarely occurs in studies of Neanderthal extinction.

Unfortunately, the ratio of ^{14}C to ^{12}C in the atmosphere has not been absolutely constant over thousands of years but varies by tiny percentages. Raw ^{14}C dates give the proportion of ^{14}C and ^{12}C in years before present (BP), in which "present" means 1950. A raw ^{14}C date may underestimate the true antiquity of a sample by as much as 10–20 percent because of these small variations in the original ^{14}C content of the atmosphere. To avoid this problem, raw dates must then be calibrated against the variations in carbon levels through time by correlating raw dates with calendar ones; this can be done by using tree rings, stalagmites, pollen sequences, or deep ocean cores, for example. Calibration standards are constantly being refined and are often indicated in a date by the appearance of the abbreviation "cal." Radiocarbon dates are routinely given with an estimate of the degree of error, such as $32,000 \pm 430$ years. High error estimates warn of uncertainty and possible degradation or contamination that survives despite the decontamination procedures.

There are two additional cautions with radiocarbon dating. First, if the sample used for dating is contaminated with modern carbon, all of the dates will be much too recent. An unnerving example cited by Wood and Higham is that a 50,000-year-old sample contaminated with only 1 percent of modern carbon will yield a date of 37,000 years old. This is a huge discrepancy if you are trying to track events that occurred on human timescales. Second, simple factors of preservation may degrade or destroy much of the collagen, the main protein in an ancient bone and the one that contains the carbon. If not enough collagen is left, the sample will have insufficient protein to indicate the true age. One way to check for such preservation problems is to look at the levels of nitrogen, a major component of collagen. If bones have very low levels of nitrogen, then the amount of collagen still left in the same sample is likely to be inadequate for dating. Directly dated Neanderthal remains from the Spanish site of El Sidrón, for example, are too low in collagen to be reliable, and dates on the bones range from 10,000 to 50,000 years ago. The huge discrepancy amply demonstrates why dating poorly preserved materials is not useful.[5] The Oxford laboratory that recently dated or redated a large number of samples led the way both in decontaminating procedures to remove modern carbon and in determining whether a sample is sufficiently well preserved to permit dating.[6] There is no reliable information to be gathered from dating a sample that is badly degraded or contaminated.

The first hint of a problem with the Iberian sites redated by Wood and Higham came from investigation of the nitrogen levels. Tellingly, only about 10 percent of the samples of 215 bones tested for redating by the Wood-Higham team were usable. Nine of the eleven sites had to be eliminated from dating because too little collagen remained. The original dates calculated on whole bone from such material—even new dates—are likely to be worthless, which means the refugium hypothesis rests on evidence from only two sites.

The two sites the team found suitable for dating were Jarama IV in central Iberia (previously dated to roughly 30,000 calendar years BP) and Cueva del Boquete de Zafarraya, which contains the youngest (most recent) date for any Neanderthal site, about 26,000 calendar years BP.

Decontamination and pretreatment methods were used in redating three Jarama IV samples and yielded dates of more than 47,000 years ago, more than 50,000 years ago, and 49,400 years ago with an error of ± 3,700 years. In other words, the new dates were generally almost 20,000 years older than previously thought and possibly much more. All three samples approached the threshold of 50,000 years, the age beyond which any radiocarbon date is suspect. The bones from Zafarraya were also much older than thought, with individual specimens yielding dates ranging from 33,300 ± 1,200 years to 46,300 ± 2,500 years. None gave dates of about 26,000 years.

The only two sites with sufficient collagen to be dated refute the hypothesis that Neanderthals retreated to the south of the Iberian peninsula and lived there until 26,000 years or so. The very notion of the last Neanderthal eking out a living on the Mediterranean shore looks very shaky. The dating team's conclusions are fairly damning. Of Jarama they say, "All radiocarbon dates on bone from the site . . . should therefore be viewed with extreme caution." Of Zafarraya they conclude, "This collection of Neanderthals should no longer be cited as providing evidence for the southern Iberian Neanderthal late refugia."[7]

This redating throws into question previous ideas about the timing of the Neanderthal extinction.[8] The obvious conclusion is that maybe Neanderthals did not live past about 40,000 years ago and that all claims that they did are based on erroneous dates.

Not surprisingly, the scientists who have worked on some of the sites that have yielded questionably young ages are skeptical of these conclusions. Clive Finlayson, a paleoecologist at the Gibraltar Museum, observed that the two redated Iberian sites were both located in very harsh, cold environments, which may have skewed the dates.[9] There is no evidence that either temperature or humidity influences the ratio of carbon isotopes, however.[10] Finlayson also argued that only two redated sites do not provide enough evidence to overturn theories supported by a number of Iberian and other sites, but the dating is crucial. The dating team has similar indications from redating of other sites like Mezmaiskaya in the northern Caucasus mountains.

Mezmaiskaya is a Russian site that included two Neanderthal skeletons. Previous dates from Mezmaiskaya suggested that the transition from the latest level bearing Neanderthal remains or tools to the earliest modern human levels occurred about 33,000 years BP.[11] This was taken as evidence of a late survival of Neanderthals in the Caucasus, similar to their hypothesized late survival in Iberia (which is no longer well supported). However, decontaminating and redating the Neanderthal skeleton that was found highest (closest to the surface) in the excavation yielded a raw date of $39,700 \pm 1,100$ BP, which calibrates to between 42,960 and 44,600 BP.[12]

Redating such key sites is not only essential but proves that, in some cases at least, paleoanthropologists have been misled by inaccurate dating.

The problem is enormous and not confined to the southern Iberian sites or to those in the Caucasus. The chronological record of much of Eurasia during the transition from Mousterian (Neanderthal) sites to Upper Paleolithic (modern human) ones is probably flawed. Higham wrote:

> Unfortunately, it is now apparent that the radiocarbon record, constructed over the last 60 years, is significantly flawed and inadequate for rigorously testing these models. This is due to the combined effects of incomplete removal of contamination and the difficulties encountered when dating samples very close to the measurement limit. . . . This was either not recognised, or not adequately addressed, at the time of dating. In addition, many of the determinations available for the Middle to Upper Palaeolithic are often only useful in the broadest chronological sense because of measurement imprecision. The development of more refined methodological approaches has had a significant effect in improving accuracy. The application of ultrafiltration for dating bone, and [pretreatment decontamination] . . . methods for dating charcoal, have shown for some sites, even those recently dated, that a large proportion of dates may be aberrant.[13]

The criteria used by Higham's team for reliable dating are extremely rigorous, as befits an absolutely essential technique.

The earliest date now postulated for the appearance of modern humans in Europe is about 36,000 years BP or 44,000 calibrated years ago.[14] Any relationship between this event and the final extinction of Neanderthals must factor in time for human populations to spread geographically and grow demographically; it is unlikely that modern humans had spread throughout Eurasia by this early date. If the last Neanderthal were reliably dated to about 40,000–42,000 calibrated years ago, then the period of overlap between modern humans and Neanderthals shrinks from about 10,000 to a few thousand years or less.

A massive redating project sampling forty key sites was undertaken by Higham and his colleagues at the Oxford lab to establish a reliable chronology for the end of the Mousterian era and therefore the timing of Neanderthal extinction.[15] Their results are vital to resolving the mysteries of Neanderthal extinctions. With stunning simplicity and outstanding accuracy, they show that the Mousterian ended across Europe between 41,030 and 39,260 cal BP with greater than 95 percent probability. Although no apparent single event or episode caused the extinctions of different Neanderthal populations, they all vanished in a fairly short period of time. However, the Mousterian did not end simultaneously across Eurasia, as it would have if extinction were the response to a single event.

Comparing the data on the end of the Mousterian era with the earliest dates for modern human arrival in Europe, there is a period of overlap of the two species between 2,600 and 5,400 years long. Allowing for the time needed for modern humans to spread across Europe into Asia, the Neanderthal extinction happened fairly rapidly after the arrival of modern humans in each area, hinting strongly that the arrival of modern humans may have been a key driver in Neanderthal extinction. And because modern humans are demonstrably a successful invasive species, it seems appropriate to turn to invasion biology to see what that approach might reveal about this extinction.

4

WHO WINS IN AN INVASION?

The outcome of an invasion depends on myriad factors.

First, of course, the species' mode of dispersal is likely to determine whether it is represented in the new habitat by small numbers of individuals on a onetime or infrequent basis or whether groups of individuals are likely to arrive together. The larger the founding group, the greater the chance that a viable population will become established in a new region. Individuals of many species are capable of undertaking considerable journeys under their own power, particularly if population pressure and the need for additional territory become pressing.

Another issue is the narrowness or specificity of the new species' needs. An example of a mammal with highly specialized needs is the koala bear. Koalas feed primarily on eucalyptus leaves, and even then, they eat only twenty of 350 known species of eucalypt. Though treetops are also essential to koalas for safety and escape from predators, koalas do not seem to care whether the treetop is in a eucalpyt or not. Obviously, a eucalypt forest would provide both food and refuge. Koalas are slow moving, very vulnerable to predators, and slow to reproduce. For example, males first reproduce at about four years and females at about two years. Females normally give birth to a single offspring per year. All of these traits work against koalas as potentially invasive species.

In contrast, a highly mobile species like the coyote, with a dietary flexibility that encompasses hunting, scavenging, and omnivory, is an able invader. Its original, pre-Columbian range was the arid prairie habitat of the southwestern United States and adjacent areas of Mexico, but this species has slowly but surely colonized the United States from coast to coast, moving north into Canada, east to the Atlantic coast, and southward to the Yucatan peninsula of Mexico. Coyotes are now common in forest, woodlands, prairies, and coastal regions as well as some major urban and suburban areas, spreading out from their original range in the arid Southwest. Not only do coyotes have more flexible dietary needs than koalas, but coyote reproductive habits are also favorable for invasive success. Coyotes are sexually mature at one year and give birth to a litter of five or six pups annually.

Dispersal is not solely dependent on an individual species' abilities, however. Plants that spread by runners may be more tightly bound to their original habitat than those with seeds that can be carried on the wind or that can be eaten and transported inside the guts of other species. Wind and water currents, massive storms, tsunamis, and dispersal facilitated by other organisms all occur; good examples are snails in mud sticking to the legs of birds, seeds that are ingested by reptiles or birds and then transported in their guts to be defecated out in a new habitat, and burs clinging to the fur of wide-ranging mammals.

One of the most frequently citied dispersal mechanisms is humans themselves, modern *Homo sapiens*. We have been moving organisms into new habitats for some 200,000 years, sometimes by accident and sometimes deliberately. An example of the extreme effects of an inadvertent introduction is the plague-carrying rats and fleas inhabiting trade goods that helped spread the Black Death through Europe in the fourteenth century, with fatal consequences for many humans. Other examples are internal parasites or lice.

Humans like to have familiar plants and animals around them in new habitats. In the last few hundred years, humans have deliberately moved large-bodied domesticates like cattle and camels around the globe. Sometimes the point of the introductions is to increase food resources, as with the elk and deer taken to New Zealand to provide hunting for European settlers, or rabbits introduced to Australia, which became a major, ongoing problem. Eighteenth-century sailors on long sea voyages often left goats on islands they encountered, in hopes that the goats would survive and serve as a food supply on future voyages. Pigs and dogs accompanied settlers of most of Oceania on their sea voyages as possible sources of food and waste removal. Similarly, the semi-domesticated canid known as the dingo was brought to Sahul (greater Australia, New Guinea, and Tasmania) by humans who originated on the Indian subcontinent.[1] However, humans had been in Greater Australia for many thousands of years before dingoes arrived.

Most texts and articles on invasion biology consider humans—if they consider them at all—to be a powerful dispersal mechanism. One of my main points is that humans should be viewed as a long-lasting and highly invasive species, one of whose main features is a strong tendency to bring "hitchhikers"—other species—along on its geographic expansions.

Did the biological attributes of modern humans make them successful invaders? They came from eastern Africa, where they had evolved from archaic humans by at least 200,000 years ago. As a primate species, these early humans were relatively large-bodied and fully bipedal. They were capable predators who made and used stone tools to enhance their ability to obtain key resources and who almost certainly made tools out of perishable wood or other plant parts.

The earliest modern humans had brains comparable in size to those of us living today. They left behind many stone tools; numerous objects of personal adornment, such as pierced shells strung on thongs or fibers; hearths; and animal bones marked with cuts from stone tools. From the very beginning of stone tools—about 2.6 million years ago—these implements were used to obtain and process animal products, including meat, fat, hide, tendon, and marrow.[2] From the very beginning or shortly thereafter, other tools were employed to procure and use a range of vegetable products, though the evidence of use of vegetable foods is more subtle because of differences in preservation between bones and plant material.

Strikingly, these earliest humans were able to hunt or scavenge animals ranging in size from hedgehogs to elephants, with the favored prey apparently being medium-sized herbivores like impalas or wildebeest. We know this because of the cut marks and unusual patterns of bone breakage on fossilized animal remains. One of the traits that make modern humans so remarkable is that they violate the rules of predator-to-prey size.

The body size of mammalian predators can be used to predict the size of prey that they favor, regardless of what zoological family the predator belongs to. Small-bodied predators favor prey weighing much less than themselves, while medium-sized predators tend to take prey up to about 66 percent of their own body size. Only large-sized predators regularly take prey larger than themselves.

The hunting style of the predator is also crucial. A predator that hunts cooperatively can and does take much larger prey than lone hunters of the same species. Thus, pack-hunting animals such as lions, spotted hyenas, and extinct predators like saber-toothed cats, for example, might be considered superpredators. Even dholes, the twenty-kilogram canids of Asia, can in large groups bring down species weighing as much as 100 kilograms. The extraordinarily wide range of species preyed on by early humans is a strong indication that they, too, were probably group hunters.

As large-bodied, highly mobile animals with a flexible diet, humans might be expected to be successful invaders, despite the fact that humans reproduce very slowly. We have a late age of sexual maturity compared with other comparable-sized mammals, and we have a strong tendency to have singletons rather than litters at intervals of several years. Neanderthals shared many of these traits. They were clearly a predatory, social, fire-building, tool-making species. Like modern humans, Neanderthals routinely hunted and killed animals much bigger than themselves.

To understand the extinction of Neanderthals and the survival of modern humans, we have to look at another very important factor: global changes in climate over time. These, too, must have had a powerful effect on who lived and who died.

The massive and unparalleled invasion of early modern humans into the rest of the world began about 130,000 years ago. Before then, early modern humans were found only in

Africa. Our close relatives, the Neanderthals, inhabited Eurasia, including the Middle Eastern region known as the Levant, but not Africa. From about 130,000 years ago, sites in the Levant show a pattern of alternating occupation by modern humans and Neanderthals that corresponds roughly with climatic changes.

Many proxies for climate can be used to trace changes through time. Ancient pollen samples indicate which types of plants were particularly abundant or rare; secondary mineral deposits such as stalagmites and stalactites, which form in caves, can indicate how much rainfall occurred during their formation; tiny planktonic sea creatures called foraminifera are preserved in ancient seafloor sediments and incorporated different isotopes of oxygen into their shells according to the ocean temperature; and fossils of micromammals such as mice, rats, shrews, or squirrels usually can live only at a limited range of temperatures, so their presence or proportions provide another indicator of climate. Synthesizing information from all of these different sources has given researchers a good framework for understanding past rainfall, temperature, and stability of climates through time. Paleoanthropologists and paleontologists use marine isotope stages (MIS) as indicators of climate. These are sometimes referred to as oxygen isotope stages (OIS) and indicate ancient temperatures through the preserved ratios of two isotopes, ^{18}O and ^{16}O. During colder periods, lighter ^{16}O evaporates preferentially, concentrating ^{18}O in the oceans and in ocean-living foraminifera. As the evaporated ^{16}O returns to earth as ice and snow, the polar ice caps become larger, making the sea levels lower. Stages with even OIS numbers were high in ^{18}O and were cold, glacial periods, whereas those with odd numbers were high in ^{16}O and were warmer, interglacial periods.

Only the last five marine isotope stages are relevant here. I will review them very briefly, stepping backward in time from the present. MIS 1 is our current climate stage, which goes back to 11,000 calendar years ago. The 11,000-year change was the end of a period colder than now that is known as the Younger Dryas. The influence of humans during MIS 1 has been so profound that a proposal has been made to call our geologic age the Anthropocene, not the Present or the Holocene, as it has traditionally been called. ("Anthropos" is Greek for "man.") The proposal is controversial, but renaming the climate stage would make an important point. Modern humans are the only living hominin during the Anthropocene.[3]

MIS 2 is the preceding stage, covering the period from 24,000 to 11,000 years ago. It is often called the Last Glacial Maximum and was the last great Ice Age in world history. Neanderthals were extinct, and only modern humans were alive, with the exception of the peculiar dwarfed hominin on the island of Flores in Indonesia known as *Homo floresiensis* or the Hobbit. This odd species and its relations to other hominins are controversial, and the last known fossils of it are 17,000 years old.[4]

Lasting from 60,000 to 24,000, MIS 3 was the period when modern humans entered Eurasia. Eurasia was also the longtime home of Neanderthals, who had flourished there for hundreds of thousands of years with many other mammals that are now extinct, including saber-toothed cats, mammoths, woolly rhinos, and cave lions. Understanding MIS 3 is crucial to understanding the extinction of the Neanderthals.

It was a period of considerable climatic instability that saw many abrupt flip-flops in climate, sometimes changing from warmer to a brief colder episode and back again in a matter of a few hundred years. Each such abrupt change within the marine isotope stages is referred to as a Heinrich Event (HE)—the most severely cold being HE 4, at about 39,300 years ago.

During MIS 3, around 39,300 years ago, a massive volcanic eruption near Naples blanketed much of central and eastern Europe with a distinctive ash, the Campanian Ignimbrite (CI).[5] A huge ash cloud covered an enormous area, depositing minute ash fragments not visible to the naked eye but which can be detected geochemically. This ash cloud undoubtedly had a major impact on temperature and on the ecosystems subjected to it. John Lowe of the Royal Holloway Institute in London and his multinational team used the volcanic ash that spewed forth in this eruption as a fixed datum in time, to assess whether Neanderthals went extinct before or after 40,000 years ago and whether modern humans appeared before or after this massive explosion. They explain: "The CI eruption was the largest within the Mediterranean area during the last [200,000 years]. . . . It liberated some 250–300 km^3 of volcanic ash, which spread over a large sector of Central and Eastern Europe; the injection of such huge amounts of ash and volatiles (including sulfurous gases) into the atmosphere is likely to have caused a volcanic winter."[6] The CI was not only widely spread but is also geochemically distinctive.

Some have speculated that the environmental impact of this eruption forced Neanderthals out of parts of Europe, leaving it open for invasion by modern humans, or imposed environmental stress that accelerated the replacement of

Neanderthals by modern humans. But the results of the careful location of the CI from northern Africa across a vast area of Europe and on into Russia reveals otherwise.

At six localities in Italy and at five more sites from northern Africa, the Balkans, and Russia, modern human sites underlie—that is, are older than—the CI. At two localities in Greece and Montenegro and one in northern Africa, Neanderthal sites also underlie the ash. The team concluded, "Our results imply that such extinction [of the Neanderthals] is likely to have occurred long before the CI eruption. . . . [Modern humans] also seem to have been widespread throughout much of Europe before the CI eruption; thus, Neanderthal and [modern human] population interactions must have occurred before . . . [40,000 years BP]."[7] Modern humans did and Neanderthals did not survive the CI eruption. The redating project at Oxford confirmed this conclusion with dated samples from many more sites. The massive volcanic eruption cannot have been the proximate cause of Neanderthal extinction because the eruption occurred on a single day or two, whereas the extinction happened over several hundred or at most a few thousand years. The CI mapping provides confirmation that modern humans were already in Eurasia at that time, demonstrating their ability to survive such an event.

The preceding and colder MIS 4 lasted from about 71,000 to 60,000 years ago. Neanderthals were present in Eurasia during this period, but modern humans were found only in Africa and the Mediterranean parts of the Middle East (the Levant) at that time.

MIS 4 was in turn preceded by the milder MIS 5 period, which began 130,000 years ago. At the beginning of MIS 5, modern humans expanded out of Africa into the Levant, marking the first time modern humans could have come into contact with Neanderthals.

Was this early expansion of the geographic range of modern humans from Africa to the Levant prompted by climate change, which altered rainfall, vegetation, and animal communities? Perhaps, and perhaps not. A more favorable climate might have enabled the African populations of early *Homo sapiens* to explore new areas, to see what lay over the next hill. In this case, what lay over the next hill if the eastern route were followed was the Levant. (It was also possible to leave Africa via Gibraltar and a western route, although the fossil and archaeological record is as yet less rich there.) Bear in mind, the geographic expansion of modern humans was very unlikely to have been a deliberate action, nor would they have known that they were leaving the African continent. On the ground, perceiving the edges of a continent is difficult; you may see a shoreline, but you do not see both coasts. Most probably, modern humans simply followed game or looked for new, unoccupied habitats at the edges of their range.

In any case, starting from the beginning of MIS 5, the only hominin remains in the Middle East were those of early modern humans, discovered at sites like Qafzeh and Skhul in Israel. A sole Neanderthal was found at level C of the Tabun site, which is roughly contemporaneous with the modern humans from Qafzeh and Skhul, but questions about the exact find spot of this fossil make it unclear whether these two species ever really coexisted in close proximity.

With the onset of MIS 4, the climate shifted again, and there was so little rainfall that formerly temperate areas became desert. Ecosystems were dramatically changed as the climate deteriorated. Early modern humans left the Levant, and Neanderthals came back in and persisted until mid-MIS 3. The last known Neanderthals in the Levant occur at Geulah Cave B, Mount Carmel, which was dated to 42,000 ± 1,700 years ago.[8] As so often with radiocarbon dates calculated years ago, the date remains controversial. Though remnant populations may have lingered in spots, Neanderthals went locally extinct in the Levant about 45,000 years ago; early modern humans returned to the Levant and spread into Eurasia. The overall picture hints that the favorable areas for living in the Levant were traded back and forth between humans and Neanderthals, with humans occupying them when the Levant was milder and Neanderthals when it was colder.

Within this framework of climate change, can we say anything more secure about the cause(s) of Neanderthal extinction?

John Shea, an archaeologist from Stony Brook University who specializes in the Levant, sees several factors interacting to produce Neanderthal extinction.[9] He characterizes the final change from Neanderthals to humans as a clear turnover event, not a gradual transition from one species to the other and one tool industry to another. Shea interprets the data as showing that, as Neanderthals waned, they were replaced by invasive modern humans who were better able to cope with the harsh conditions.

Shea has estimated that the worsening and unstable climate during MIS 3 caused the Mediterranean woodland habitat of Neanderthals to shrink by perhaps 75 percent while open, cold steppelands increased. Loss of continuous, suitable habitat may have left Neanderthals with insufficient resources for survival of viable populations. Shea's results are paralleled by other studies. If, as Clive Finlayson has suggested, Neanderthals were ambush hunters depending on vegetation to conceal them as they crept up on their prey, this would make excellent sense. A lack of vegetation cover might form an enormous disadvantage to Neanderthals, particularly if—as anatomical studies by Steven Churchill of Duke University indicate—Neanderthals killed animals not by throwing spears or shooting arrows but by stabbing their prey with a spear and wrestling the animal down. Such a technique can be described as close-contact hunting. It was unquestionably dangerous and must have required enormous strength and skill.

If climate changes made it harder for Neanderthals to succeed following their usual mode of hunting, would this have been enough to push them into extinction? Perhaps. But it is also relevant that the severe bout of climatic deterioration that occurred during MIS 3 at about 45,000 years ago coincided with the reinvasion of the Levant by modern humans, who may simply have dominated, displaced, or outcompeted Neanderthals.

Shea argues convincingly that the use of complex, projectile weapons, such as bows and arrows or spears propelled by atlatls (spear throwers), offered modern humans a substantial edge. Such weaponry is known from North, South, and East Africa—the indigenous homeland of *Homo sapiens*—well before 50,000 years ago. New finds from South Africa dated to 71,000 years ago push the origins of such complex weaponry back even further.[10] Projectile weaponry is not, however, associated with Neanderthal remains or archaeological sites. Shea writes:

Unlike hand-cast spears and non-piercing weapons (boomerangs, throwing sticks, etc.), projectile weapons are light, allowing a single hunter to carry many of them at the same time. They fly quickly, allowing them to be used on small, fast-moving targets as well as larger stationary ones. They retain energy longer in flight, allowing them to be used against larger dangerous prey, or other humans, with less risk of injury. . . . In a word, projectile weaponry is niche-broadening technology. It underwrites one of the most distinctive derived features of Eurasian (indeed all) human adaptations after [50,000 years], our broad and flexible ecological niche. Like no other subsistence adaptation, complex projectile technology makes *Homo sapiens* the quintessential ecological generalist, and in evolutionary competition, generalists always beat specialists.[11]

Early modern humans invaded the Eurasian heartland not long before the last Neanderthal died. To establish the very first appearance of modern humans in Eurasia, neither stone tools nor ancient hearths help us much with chronology. Neanderthals lived all over Europe before modern humans got there and made their hearths, dropped their tools, and left their dead in many areas. After modern humans arrived, they too did the same, leaving tangible evidence of areas where they lived or simply passed through.

Because of the inherent uncertainty of linking tool industries to a particular hominin species, only fossils, and good ones in well-dated contexts, will do to anchor a chronological framework for who lived where and when. One approach is to date bones from archaeological sites with a distinctive stone tool industry—and to select for dating those bones that bear clear evidence of burning, cutting with stone tools, or other kinds of artificial shaping. Even better, but more difficult, is the dating of the skeletal remains that can be clearly identified as belonging to modern humans themselves. Both approaches were used by the Oxford team to establish the end of the Mousterian industry.

Why is directly dating skeletal remains difficult? First of all, not all parts of the skeleton can be definitively identified as Neanderthal or modern human. Second, there are simply not as many as one would hope for. Third, some curators are reluctant to allow researchers to take (and destroy) a sample from a precious specimen, even though the amount of bone needed is very, very small (about one gram). Only relatively few hominin remains have been dated directly using modern techniques, and fewer still would constitute a "first" or a "last" appearance.

One is a diagnostic skull of a modern human that is dated to between 46,000 and 63,000 years BP (uncalibrated) from Tam Pa Ling, Laos, which was first announced in 2012.[12] If modern humans were in Laos more than 45,000 years BP, they must have traversed central Europe even earlier. However, the radiocarbon date was carried out on small charcoal samples and is very near the limit of ^{14}C dating. Other methods of dating were applied to soil and sediment samples from Tam Pa Ling. These confirm the general antiquity of the specimens but do not refine the accuracy of the date, which is poor. Also, combining all dates from all techniques with a stratigraphic section shows that some dates are out of sequence, which means either that stratigraphic layers are mixed or that some samples may be contaminated and yielding false dates. Doubts have been raised about the antiquity of this specimen.[13]

Another indicator of early hominin presence in China comes from the recent discovery of two teeth from Lunadong Cave in China.[14] They are certainly hominin teeth and the authors believe they belong to modern humans, not Neanderthals. Assigning isolated teeth to either species is difficult, however. If the teeth are modern, then their position in the ground—they were excavated from between two geological layers dated to 129,000 and 70,000 years ago—suggests modern humans may have arrived in China earlier than expected, based on previous dates. The remains are too old to be dated by radiocarbon techniques and cannot be directly dated.

A stronger case can be made for the partial human jaw from Kent's Cavern, England, dated to 45,000 and 41,000 calendar years BP after ultrafiltration and decontamination. To get from the Levant to England, modern humans must have passed through parts of mainland Eurasia first. This dating matches nicely with the earliest Aurignacian (modern human) sites in Europe, which are about 42,500 calendar years BP.

The Kent's Cavern jaw is essentially contemporaneous with two isolated milk teeth from the Grotta del Cavallo in Italy, dated to 45,000 and 43,000 calendar years BP.[15] Importantly, the Grotta del Cavallo teeth are associated with a "transitional" stone tool industry known as the Uluzzian that was, until recently, attributed to Neanderthals. However, redating the teeth from Grotta del Cavallo, and demonstrating that they are from a modern human, refutes the link between the

Uluzzian and Neanderthals. It also demonstrates an overlap of Neanderthals and modern humans in Italy of possibly 3,000 years.

The fourth especially significant fossil is a jaw from the Romanian site of Peştera cu Oase, dated to about 46,000 years ago. It is not associated with any tools, but the antiquity of the site is generally confirmed by its position below (older than) the 39,300-year-old CI.[16]

A final (at present) line of evidence comes from the peopling of Australia. As I said earlier, no hominins of any type lived in Australia until early modern humans arrived about 50,000–45,000 years ago, so stone tools, hearths, and other types of archaeological evidence are more reliable indicators of modern human presence in Australia than in Eurasia. Sites like Devil's Lair, Lake Mungo, and Naworla are dated to between 48,000 and 44,000. If modern humans arrived in Australia at almost 50,000 years ago—and this is a conservative date—then they must have gotten there through Eurasia. Somehow. But Australia was not connected to the Asian mainland at this time, and the minimum distance of travel over open seas required to reach Australia was eighty kilometers. Many Australian scholars argue that an ability to reach Australia at that time indicates that these humans had boats and considerable knowledge of seafaring, indicating some sophisticated skills. Surely such "advanced" and "modern" people were better hunters than Neanderthals. Or is that simply hubris talking?

The big question still remains: Was the invasion of modern human populations the factor that pushed Neanderthals into extinction, or was the primary driver the changing climate?

5

HOW DO YOU KNOW WHAT YOU THINK YOU KNOW?

It is nearly axiomatic in science that a hypothesis is only as good as its ability to be tested against real data. Ideas that are exciting, intuitively appealing, or politically correct are not worth much if no data can possibly be gathered to see whether those ideas are actually true. Half of the struggle in science is to figure out what the predictions of a hypothesis are and how you would determine whether those predictions are true. In historical sciences like paleoanthropology, the situation is complicated by the fact that no experiments can be carried out; evolutionary history cannot be rerun like a recording of a favorite movie, nor can the plot be replayed with a few variables changed.

There are currently two main contenders for the explanation of Neanderthal extinction. The first is that the climate instability and changes during MIS 3, the period during which extinction occurred, drove Neanderthals to extinction. Clive Finlayson is one of the foremost proponents of the climate change hypothesis. Finlayson has helped compile habitat maps of western Europe that show the vegetation types and animal ecosystems through time. He and others have maintained that there is no direct evidence that Neanderthals and modern humans ever coexisted in exactly the same places at the same time—this would be an extremely difficult thing to prove—though they shared a general region. He proposes that each species expanded or contracted its geographic range as climate changed. "It was a shifting balance," he says, "A sort of semi-permanent geographic coexistence."[1] And when modern humans reached the Eurasian plains, where Neanderthals were already rare if not locally extinct, the question became moot. There were neither geographic barriers nor competitive barriers to the expansion of moderns once Neanderthals had retreated to an ever-dwindling Mediterranean area from which they were never able to recolonize the central and northern areas again.

Finlayson emphasizes that such a dynamic coexistence was dependent on the carrying capacity of the habitat. *Carrying capacity* is an ecological term referring to how many organisms a particular habitat can sustain indefinitely. It is a useful concept but a difficult one to measure or estimate precisely. Finlayson suggests that the Eurasian ecosystem was not entirely "full"—not so heavily populated by mammals that no new species could be supported—and so Neanderthals and modern humans were not in direct competition with each other. He believes their lifestyles involved complementary adaptions. Neanderthals, he argues, were ambush predators that used the cover of brush and tall grasses as they stalked their prey, until they got close enough to kill with handheld weapons. If this interpretation is correct, then it is incredibly impressive that Neanderthals managed to kill large prey. Modern humans had long-distance weapons—projectiles—and favored the vast open plains of the steppeland and tundra.

Both of these assumptions are supported by archaeological and anatomical evidence.

"Inter-specific (or inter-population) competition is a very difficult phenomenon to demonstrate in any extant wild population today," observes Finlayson, adding, "It is practically impossible to know whether or not Neanderthals and Moderns [*sic*] were in competition."[2] Looking at the carefully reconstructed mosaic of habitats and ecosystems through MIS 3, Finlayson sees "a pattern of extinction that is related to bioclimatic zones, strongly suggesting that a climate-driven impact was responsible."[3] In seeking an explanation for Neanderthal extinction, it would certainly be unwise to ignore or dismiss the abundance of evidence that the world climate underwent multiple fluctuations, including one that occurred about 40,000 years ago and was probably related to the Campanian Ignimbrite. But there have been many such fluctuations, just as severe, none of which pushed Neanderthals into extinction. Why would that particular cold spell wipe out a species that had been successful for hundreds of thousands of years?

Extinction by climate change is as complex a topic as extinction by competition. Ecologist Abigail Cahill of Stony Brook University and colleagues note, "Our goal is to understand which proximate factors cause extinctions due to climate change. However, we first need to establish that such extinctions are presently occurring. Few global species extinctions are thought to have been caused by climate change. For example, only 20 of 864 species extinctions are considered by the International Union for Conservation of Nature (IUCN) . . . to potentially be the result of climate change, either wholly or in part . . . and the evidence linking them to climate change is typically very tenuous."[4] This investigation into how climate change actually causes extinction uncovered some nuances. Of the twenty extinctions related to climate change by the IUCN, only one involves a mammal—an island rodent, *Geocapromys thoracatus*. Clearly, though climate change is not a common or frequent cause of mammalian extinction today, that finding does not mean that climate change had no role in Neanderthal extinction.

The same group reviewed 136 studies focusing on local extinctions and climate change, to look at whether populations sitting at the edge of their physiologically determined temperature tolerance were unusually vulnerable to extinction. Only seven studies linked local extinction to climate change. Of those, only two involved mammals: the pika, *Ochotona princeps*, which has a limited tolerance to both high and low temperature extremes; and the desert bighorn sheep, *Ovis canadensis*, the extinction of which is attributed to decreased rainfall and subsequent changes in vegetation in its range. The other climate-caused extinctions involved fish, planaria, butterflies, and lizards, which are probably poor analogues for Neanderthals.

Even natural and potentially more major changes in climate are not a common cause of extinction. Out of the 136 studies Cahill and colleagues reviewed, only four identified extinction as being related to natural climatic oscillations known as the El Niño Southern Oscillations. These studies involved butterfly fish, the toad *Bufo boreas*, corals, and fig wasps in Borneo—again, organisms not very similar to hominins. Thus, making the case for climate-caused extinction of Neanderthals is likely to be as difficult as making a case for invasion-caused extinction—and yet, there is good evidence for both.

Finlayson is adamant that Neanderthals did not die out because they were stupid, inept, or in some way inferior to modern humans. He writes in one of his books, "A popular thesis is that competition from the newly arrived and 'superior' Moderns [*sic*] caused the extinction of the Neanderthals. The only basis for the argument is that of an apparent association between the arrival of the moderns into Europe and the extinction of the Neanderthal."[5] Nonetheless, that chronological coincidence is important—and, as I will show, is only one line of evidence implicating the modern human invasion in Neanderthal extinction.

Finlayson is certainly correct that there has been a tendency to stereotype Neanderthals disparagingly as brutish cavemen incapable of sophisticated thought or behavior since their fossils were first discovered.[6] Yet in recent years, many behaviors long thought to indicate intellectual modernity have been shown to exist among Neanderthals. Those behaviors include use of ochre and other pigments, the wearing of objects of personal adornment or jewelry—which were possibly badges of membership in different clans, families, or groups—and the use of marine resources and even birds. All of these were once thought to be the exclusive purview of modern humans.

Curtis Marean of Arizona State University challenges the "laundry list" approach to defining modern human behavior, calling it simplistic. It is not a specific behavior or list of behaviors that made modern humans modern, he suggests, but an overall flexibility and ability to cope. Marean expresses his view of the differences between modern humans and Neanderthals: "Neanderthals had a Neanderthal way of doing things, and it was great as long as they weren't competing with moderns," he says. Modern humans who invaded Europe found a different ecosystem than that in which they had evolved in Africa, but they invaded successfully nonetheless. As climate changed and competition with Neanderthals increased, modern humans may have been more adaptable and flexible in their behavior. "The key difference is that Neanderthals were just not as advanced cognitively as modern humans," Marean reasons. He is currently exploring the idea that systematic marine exploitation among early modern humans promoted prosociality—a firm bonding or identification with a group that resulted in altruistic behavior, cooperation, and enhanced communication that was lacking among Neanderthals.[7]

Is this a slur against our Neanderthal ancestors? Or simply the truth?

If climate change doesn't account fully for Neanderthal extinction, can we demonstrate competition between modern humans and Neanderthals? And what exactly is meant by "competition" in this context?

The key to competition lies in a finite or limited set of resources shared by two species that is insufficient for both. This key concept is often called Gause's law of competitive exclusion. It predicts that, over time, one species will drive its competitor away or to extinction. Not only the resources themselves but often the way they are used is invoked.[8] The classical concept of interspecific competition was originally developed by statistician Alfred Lotka in 1910, elaborated by Italian mathematician Vita Volterra in 1926, and then reframed in 1960 by Russian ecologist Georgy Gause.[9] Predicting which species will "win" is formalized as a series of equations dealing with population sizes of both species, the carrying capacity of the ecosystem, how fast those populations grow, the duration of competition, and the extent of competition between the two species. To keep the model fairly simple, Lotka and Volterra considered relatively few factors that might influence whether two species could coexist. What they looked at was a zero-sum game with limited resources and a single immigration event. The experiments on which the model is based were carried out in water containers with two species of *Paramecium,* tiny protozoans. The Lotka-Volterra equations both simplify and clarify the complexities of reality. They show that the only circumstance in which two highly similar species can coexist occurs when the competition *among* members of one species has a greater effect than competition *between* the two species. In every other case, competitive exclusion takes place.

However, as ecologists John Terborgh of Duke University, Robert Holt of the University of Florida, and James Estes of the University of California, Santa Cruz, have written, "No ecologist, not even the wooliest theoretician, believes that the Lotka-Volterra model literally describes all the rich behavior of any actual ecosystem."[10] Though competition can be difficult to prove in the paleontological record, as Clive Finlayson observes, there are indicators of competition and its severity that can be detected. One of the most obvious and easily tracked forms of competition is for food. This does not mean something mundane, like "don't steal my tomatoes," but something that is biologically connected to almost everything about a species.

How a species obtains food will influence not only its diet but often its preferred habitat, its way of moving about the world (locomotion), and its reproductive patterns, down to details like number of offspring per year, season of reproduction, and how young are cared for. Foods that are distributed in clumps, like fruiting trees, tend to be used by social species, because the amount of food at any one time is enough for a social group. Marean would include shellfish in the category of food that occurs in clusters sufficient to feed a group, which promotes social interactions. In turn, sociality may enhance safety from predators, for example. Animals that eat scarcer foods or ones that occur in smaller units are likelier to be loners.

Another key attribute that influences species behavior has to do with patterns of reproduction, growth, and development in the species. Ecologists often distinguish two types of species. One reproduces slowly at long intervals, having few offspring at a time, which are immature at birth and thus require a high degree of parental care during development. These are so-called K-selected species, named for the mathematical symbol for the carrying capacity of an environment, K. Learning is important in K-selected species, starting from birth. K-selected species are often born in protected areas or dens, where they live and grow for some time before venturing into the outside world when they are able to behave like the adults of their species. Both carnivores and primates are often K-selected species, and typically they are strong competitors for limited resources.

The other extreme is an r-selected species, named for r, which is the symbol of the rate of growth of a population. An r-selected species reproduces rapidly and bears many offspring that, at birth, are able to move, eat, communicate, and behave much like adults.[11] A good example would be a wildebeest, which can run and feed minutes after birth. Although there is good evidence that Neanderthals matured more rapidly than modern humans, both are *K*-selected species that invested heavily in single offspring who required a lot of parental investment.[12] Thus at this fundamental level, both Neanderthals and early humans were the same general type of animal.

Obtaining food also dictates a great deal about a species' interactions with other species—friendly or deadly—and an animal's ability to withstand unpredictable fluctuations in climate, habitat, or rainfall. Many species have a few primary food sources and also fallback foods, used when the primary foods are scarce or in unfavorable seasons. Ecologists debate whether the primary food or the fallback food is likely to have more influence on a species' anatomy and behavior.

It is often said, "you are what you eat," but this truism extends far deeper into a species' biology than is usually appreciated. To understand Neanderthal extinction and modern human survival, we need to look at what was on the menu.

6

WHAT'S FOR DINNER?

Food competition is one of the most fundamental and powerful forces in an ecosystem. Charles Elton, the founder of invasion biology, conceived of the basic organization of ecosystems as a trophic or dietary pyramid.

At the bottom or fundamental step, energy comes to terrestrial ecosystems in the form of sunlight, the basic "food" of primary producers: plants. Plants are primary producers that tend to be abundant, with many individuals in a species. Their main limitations are access to sunlight, water, and space.

The next step on the trophic pyramid is filled by primary consumers. These are the herbivores, which are sometimes subdivided into grazers (grass eaters), browsers (leaf eaters), and frugivores. These species consume the plants or primary producers. So as not to run out of food, and to allow the plants to regenerate, there must be far fewer primary consumers than primary producers. In fact, roughly 90 percent of the energy that comes into the ecosystem from sunlight goes into making plants, with only about 10 percent of that energy available to primary consumers.

At the apex of most ecosystems, on the third step up the trophic pyramid, is an even rarer type of organism: the secondary consumer. These are the insectivores (insect eaters) and the carnivores (meat eaters). Among mammals, lions, hyenas, jaguars, wolves, and other carnivores fulfill this trophic role because they eat the primary consumers. Once again, only 10 percent of the energy from the previous level, the primary consumers, is available to secondary consumers. Sometimes secondary consumers or carnivores are divided into large carnivores, those bigger than fifteen kilograms, and mesocarnivores, smaller species whose diet may be composed of up to 60 percent nonmeat foods, such as insects, fungi, plants, or fruit. A coyote is a good example of a mesocarnivore.[1]

To demonstrate competition between two species, biologists routinely look at their trophic level. Obtaining food is not only a lifelong task, but its demands dictate a great deal about a species. In 1904, the naturalist Joseph Grinnell, first director of the Museum of Vertebrate Zoology at the University of California, Berkeley, observed, "Every animal tends to increase at a geometric ratio, and is checked only by limit of food supply. It is only by adaptations to different sorts of food, or modes of food getting, that more than one species can occupy the same locality. Two species of approximately the same food habits are not likely to remain long enough evenly balanced in numbers in the same region. One will crowd out the other."[2] One of the terms often used by ecologists is that two species compete if they are members of the same *guild*, in an analogy to the medieval trade organizations, such as bricklayers or weavers. A weaver does not compete with a bricklayer, but he or she will compete with another member of the same guild. In terms of the fossil record, a guild might be that of large-bodied predators or, in reference to their prey, a grazer or browser.

With many mammals, membership in a guild or dietary group can usually be deduced from anatomy. Predators have strong teeth and jaws, often filled with sharp slicing teeth. They may be swift pursuit predators, like wolves; or agile stalkers and pouncers on prey, like lions or saber-toothed cats. Detailed anatomical resemblances usually show incontrovertibly that an extinct species was (or was not) some sort of cat or felid, a canid or doglike animal, a hyaenid like spotted hyenas, an ursid like a grizzly bear, a skunklike or mustelid animal, such as a wolverine. These are the classical zoological families in the order Carnivora, informally called carnivores. Further refinements in the diets of carnivores can be deduced from features like the size and proportions of various teeth and the adaptations of the limbs.[3]

Although modern humans belong in the order Primates, not the Carnivora, and do not possess the slicing or crushing teeth, strong jaws, powerful limbs, and sensory abilities that most carnivores do, behaviorally *Homo sapiens* is unquestionably predatory. Our ability as predators hinges on our ability to make and use tools to perform the activities for which carnivores use their jaws, senses, legs, teeth, and claws. Tool using is a basic attribute of our lineage that began long before *Homo sapiens* evolved. Abundant evidence proves that our lineage and many of our near relatives' have been predatory from the first appearance of recognizable stone tools.

The archaeological record of tool making and tool using by Neanderthals and modern humans is rich. We know from hundreds of thousands of deliberately manufactured and used stone tools and hundreds of thousands of cut-marked, deliberately broken, and sometimes burned remains of prey species that both Neanderthals and modern humans were adept at making tools and were skilled at hunting large game.[4]

Analysis of the fauna preserved at different sites indicates a substantial overlap in the animal-based portion of the diets of these hominins, with a heavy reliance on wild horses, red deer (elk), reindeer, and wild cattle like the aurochs and bison, all of which are depicted in the exquisite cave paintings created by modern humans in places such as the Lascaux and Chauvet Caves in France or Altamira in Spain, long after Neanderthals were extinct. At several sites, studies comparing the fossilized bones of prey species killed by Neanderthals and of those killed by modern humans in the same area show no statistical differences.[5] In other words, each hominin species apparently ate what prey was there and what was abundant. While they shared a geographic region, they shared—and competed for—the prey species that were also there.[6]

But what about plant foods? Fossilized remains of plants are far less common than those of animals; leaves, fruits, or even seeds only preserve well in rare circumstances. Fortunately, the relative contribution of plant and animal resources to a species' diet can also be measured by the isotopic composition of the hominins' bones, because the atoms that comprise the food an individual eats are incorporated into the individual's bone and teeth. As both carbon and nitrogen are present in all organisms, the different isotopes of each element are very useful for this purpose. The proportions of ^{12}C and ^{13}C and of ^{14}N and ^{15}N can be recovered from bones and used to distinguish grazers (grass eaters) from browsers (leaf eaters) and primary from secondary consumers. In aquatic ecosystems, diatoms and other planktonic forms play the role of primary consumer analogous to that of plants in terrestrial ecosystems.

Isotopic composition must be measured on a site-by-site basis, because soil chemistry, rainfall, and other factors influence the values derived from analysis of the hominin or other animal bones. It is standard practice to analyze species of known diet—such as moose for a browser and a wolf for a carnivore—from the same sites as the hominins in order to calibrate the isotopic differences that are incorporated into bones for each particular ecosystem, soil type, and region.

The thirteen isotopic studies of Neanderthals and fourteen roughly comparable studies of early modern humans are summarized by Michael Richards of the Max Planck Institute of Evolutionary Anthropology and Neanderthal expert Erik Trinkaus of Washington University in St. Louis.[7] The Neanderthal studies yielded remarkably consistent results. Investigations of adult Neanderthals ranging in antiquity from about 120,000 to 37,000 years BP (uncalibrated) by several different teams of researchers say the same thing: Neanderthals' dietary protein came primarily from large terrestrial mammals, like the wild horses, elk, reindeer, and aurochs that dominate their archaeological sites. The values for both carbon and nitrogen isotopes were very similar to those of the top carnivores in those areas, such as cave lions, wolves, and hyenas. There is no evidence from isotope studies that Neanderthals ate a lot of marine foods. Over a long period of their existence—including the period that saw the dwindling of their populations—Neanderthals appeared to have a very stable diet and trophic adaptation.

There are also fourteen studies of early modern humans from Europe, though only ten of these report both carbon and nitrogen isotopes. The comparison is not perfect. Only one, from Peştera cu Oase in Romania, is directly dated to a period that overlaps with Neanderthals at about 43,000 years BP (uncalibrated). The other samples are from modern humans after local extirpation or even after global Neanderthal extinction.

Most of the isotope studies of the modern humans also indicate a dependence on large terrestrial prey, but the dietary indications are more variable. The Peştera cu Oase human was unquestionably a top predator, with values indicating that this individual competed with wolves and hyenas, which relied heavily on red deer for their prey. The isotope studies from three sites in particular—Peştera cu Oase, Arene Candid IP in Italy, and La Rochette I in France—also show values that suggest reliance on aquatic resources like fish. Faunal analyses indicate that modern humans took a broader range of prey, including small animals and mollusks, than Neanderthals.[8] Whether the broader dietary spectrum of modern humans reflected superior intelligence, population pressures, or simply technological differences, both modern humans and Neanderthals were apex predators with a protein-rich diet.

In 2006, Hervé Bocherens and Dorothée Drucker of the University of Tübingen summarized the then-available isotopic information on the diets of modern humans and Neanderthals. They also concluded that Neanderthals and modern humans had a very similar diet, "based primarily on proteins from open environment herbivores. In areas where they coexisted, both hominids would have been in direct dietary competition."[9] In fact, they detected a trophic rigidity or a lack of variability among Neanderthal diets, saying, "The isotopic data available for Neanderthals suggests that no significant dietary change occurred between individuals dwelling under more forested and more open environments: a subsistence pattern based on open environment herbivores continues even under forested conditions."[10] In diet, both Neanderthals and modern humans differed from the cave bear and brown bear, which coexisted with them during this time and which were more omnivorous. However, a more recent review by Bocherens reveals some new subtleties about Neanderthal diets over time and space.[11]

Bocherens synthesized isotopic studies from a group of sites in western France and two in Belgium. During MIS 3, both regions had a cold-adapted fauna featuring a group of mammals that is sometimes referred to as the mammoth steppe fauna. It included mammoths, woolly rhinos, elk, horses, reindeer, and large aurochs (primitive undomesticated cattle), and bison here collectively called bovids. The mammoths and bovids apparently fed on the open-country grasses, which gave those species a high ^{15}N value and a negative ^{13}C value. Horses ate plants from open habitats and closed forests. Reindeer, a classic cold-habitat and open-country species, ate different plants from those consumed by the horses, bison, and deer. Reindeer ingested lots of lichen, which gave them a more positive ^{13}C value and a less positive value of ^{15}N.

Comparing isotopic ratios in Neanderthal and hyena bones from the same set of sites pointed to much more use of mammoth and rhinoceros by Neanderthals as compared with the species eaten by hyenas, which relied more on reindeer. Bocherens argues that the mammoth and rhinoceros must have been hunted by Neanderthals, reasoning that if these mammals were scavenged from natural deaths, hyenas would have had an equal or better chance than humans of scavenging their remains. Bocherens also suggests that the carnivores may have handled the competition from Neanderthals by favoring the smaller herbivores as prey, leaving Neanderthals to cope with the megaherbivores like mammoths and rhinos.

In a still more recent synthesis of isotopic data, four French colleagues—Virginie Fabre and Sylvana Condemi, both of the Centre National de la Recherche Scientifique at the Université de la Mediterannée, and Anna Degioanni and Estelle Herrscher, of the Laboratoire Méditerranéen de Préhistoire Europe Afrique—came to a conclusion similar to that reached by Bocherens. As did previous researchers, they detected a consistent reliance on a large proportion of meat in Neanderthal diet over time, notwithstanding climatic or ecological fluctuations—which quite possibly suggests a deleterious inflexibility (referred to as "a very stable trophic adaptation").[12]

In 2012, D. C. Salazar-García, then a graduate student at the University of Valencia, carried out an isotopic study of four hominin remains from various Spanish Neanderthal sites. He focused in part on MIS 3 sites that were located in regions with a much milder climate than those studied by others in central and western Europe. Because of availability of potential foods such as seeds, acorns, wild berries, and wild olives in the milder Iberian climate, he postulated there might have been a greater reliance on plant or marine foods in the southern part of Iberia than in the colder sites. Such a finding would reinforce the idea that Neanderthals retreated southward to find a more temperate climate. Nonetheless, on the basis of identified isotopes, Salazar-García reported a dietary reconstruction that was very similar to that found elsewhere, marked by heavy dependence on large- and medium-sized terrestrial prey and no evidence of heavy use of plant foods or smaller game.[13] The same trophic rigidity found at more northerly sites showed up at the Spanish sites. And although microscopic examination of dental calculus from Neanderthal teeth and of the edges of stone tools at the sites showed some evidence of plant processing and consumption, the isotopes did not, indicating little reliance on such foods.

Salazar-García's isotopic work conflicts with the conclusions based on faunal and taphonomic analysis of Vanguard Cave and Gorham's Cave in Gibraltar, by a group led by Chris Stringer and Clive Finlayson.[14] The discrepancy sent me back to the original papers, which I thought I knew well because I previously had written a commentary on one of them.[15] I will summarize here some of the key information reported in the article, including those points to which I simply did not pay enough attention at the time.

At Vanguard Cave, remains of bottlenose and common dolphin, fish, and monk seal—two seal bones bearing cut marks—were excavated together with the remains of 149 mollusks, several hearths, and Mousterian tools. There were no Neanderthal bones, however. Perhaps because the use of marine resources was relatively unexpected at a Neanderthal site, the nine bones from marine mammals and the 149 mollusks assumed greater importance than the remains of terrestrial mammals. Although the marine mammals were present, they were sparsely represented by only about 4 percent of the bony specimens compared with terrestrial mammals, 86 percent of which were herbivores and 10 percent of which were carnivores. The majority of the bones from terrestrial species were those of the Iberian wild goat, *Capra ibex* (121 specimens), a common prey of Neanderthals that dominated the assemblage overwhelmingly.

The 149 mollusks were clustered in a single layer of twelve square meters, along with a hearth, knapping debris, and Mousterian tools. Most of these were mussels, *Mytilis gallo-provincialis*, which probably lived in a nearby estuary. In numbers, the mussels were relatively few for the size of the excavation, compared with numbers at later sites where shellfish were frequently and systematically exploited. In other words, the Vanguard Cave remains do not suggest a heavy or systematic reliance on marine resources but only an occasional use, a point I did not fully appreciate until well after I wrote my commentary.[16]

The analysis of Gorham's Cave, which is literally within sight of Vanguard Cave, was published along with data from Vanguard. Here, too, I overlooked part of the import of the analysis. Gorham's Cave, like Vanguard, bore no hominin remains but only tools and animal remains. Of the 226 mammalian specimens that would be identified from the Mousterian level at Gorham's Cave, a single bone (less than 0.1 percent) was from a seal, and the rest were from terrestrial mammals. The most commonly represented mammals were leporids: rabbits comprise 140 specimens, or 62 percent of the fauna. At the Aurignacian or modern human level within Gorham's Cave, 1,026 specimens could be identified. Fully 723 specimens (70 percent) were rabbits, 139 specimens (14 percent) were goats, and one specimen was a seal. As the authors remark, there is no significant difference in the faunal representation in the Neanderthal and modern human levels—and both, as I should have appreciated, showed a strong dependence on terrestrial animals. The authors are an impressive group of paleoanthropologists, who claimed that these sites demonstrated a systematic and repeated practice by Neanderthals of using marine resources. I no longer agree that this conclusion is amply warranted. The marine resources are very few compared with the terrestrial ones.

Another point about the results from the Gibraltar caves that I overlooked—embarrassingly as it involves one of my research specialties—is the extraordinary frequency of cut marks or other hominin-caused modifications on the bones. Almost half of the Vanguard Cave bones show cut marks or other indicators of human activity. Intuitively, this percentage is much higher (by about a factor of ten) than any I have seen in the assemblages I have analyzed. The high frequency of marks suggest a particularly intensive processing of the remains to extract every last scrap of edible food. It is, of course, difficult to assess what a normal level of cut marks and impact fractures might be, as such numbers vary with the animals being butchered, the skill and tools of the butcher, and what the objective of the butchery is, from complete removal of all meat, to removal of the hide, or harvesting of tendons or bones for other uses.[17]

Yet these highly processed finds are hauntingly evocative, suggestive of a population making do on dwindling resources. Most critically, the exploitation of marine resources at these caves by Neanderthals is not confirmed by finds from other sites or from isotopic analyses from sites in southern Iberia, where potential resources were similar.

In truth, hunting big game is a risky and unpredictable strategy for procuring food, yet both Neanderthals and early modern humans relied heavily on hunting. Mary Stiner and Steven Kuhn of the University of Arizona compared the yield in terms of calories and nutrients of different types of food relative to the costs of obtaining and processing those foods.

They comment, thoughtfully, "Large game animals can yield high average return rates but these resources are unpredictable as staple food sources. . . . In [living] humans, the high day-to-day variance in protein and fats available to children and pregnant or lactating women will limit the reproductive potential of a highly carnivorous population. While the long existence of Middle Paleolithic lifeways across the Old World indicates that adult females in these societies enjoyed reasonable levels of reproductive success, women's fertility would have remained very low due to the unpredictable diet and the necessity of women's cooperation and ready proximity for hunting operations. Middle Paleolithic populations seldom attained large sizes and were subject to frequent crashes."[18] They suggest that what made a large difference between Neanderthal and modern human success in the Eurasian ecosystem was that Neanderthals apparently had fewer means of evening out variations in the food supply compared with those practiced by modern humans. Another way of looking at this is that Neanderthals had a more limited range of fallback foods than modern humans.

Despite a possible overemphasis on the marine resources in the faunal remains, until recently Gorham's Cave offered some of the best evidence of the late survival of Neanderthals in Mediterranean climates (to about 28,000 years BP)—that is, if the cave is accurately dated. But in the recent study by Higham's team, the collagen levels in the Gorham's Cave bones were found to be too low to yield a reliable date. Additional samples from Gorham's Cave or Vanguard Cave, or possibly from new sites, may clarify whether this area was used as a last refuge by Neanderthals.[19] As of now, the dating evidence does not support or refute this interpretation because the bones cannot be dated accurately using the best available techniques. Once again, dating is a crucial issue. Until it is resolved, it is difficult to know whether the evidence of un-usual dietary breadth and late survival of Neanderthals in the south is real. The importance of dietary breadth is that using low-quality but easily obtained prey was an effective way to compensate for gaps in the supply of high-quality food items, like large game.

Can we say that the isotopic studies reveal everything about Neanderthal and modern human diets? No; some dietary components are probably invisible, as plant foods often are from archaeological sites. In general, the isotope studies agree with findings based on the preservation of bones of different species in various sites. However, the isotopic values suggest a higher proportion of megaherbivores in Neanderthal diets than do the faunal analyses. This may be a consequence of the so-called schlepp effect. The sheer difficulty of transporting parts of extremely large animals—to carry something awkward and cumbersome is referred to as *schlepping* in Yiddish—may have caused hunters to field dress the carcasses rather than try to carry large, meat-laden bones or drag enormous car-casses back to the living area or camp intact. Not only the nuisance value of transporting large carcasses but also the inherent danger of remaining in the area of the kill site when carnivore competitors appear might discourage such behavior. Any intelligent Neanderthal would have stripped the meat from the bones at the kill site and carried only the meat back to their campsites, causing the discrepancy between faunal remains and isotopic readings.

Nonetheless, at least one site provides reasonable evidence of Neanderthals' abilities to hunt megaherbivores, or animals of extremely large size, like mammoths and woolly rhinos. The Belgian site of Spy (which has been redated recently to about 36,000 years BP uncalibrated or almost 40,000 calibrated) has yielded two adult and one juvenile Neanderthal remains, as well as many dental and crania parts of mammoths, many individual horses, hyenas (the second most abundant species), and fewer remains of woolly rhino, large bovids, and reindeer. There are also uncommon bones from cave bear, cave lion, and wolf.[20] The skeletal remains of the mammoths are highly skewed compared with what would be expected if whole animals were killed at the spot; there are no limb bones, vertebrae, or ribs. Because mammoths were mammoth in the other sense of the word, a kill site should include all the skeletal elements of these animals. In contrast, at Spy, the mammoth remains are nearly all derived from the heads. Belgian paleontologist Mietje Germonpré and her colleagues suggest that possibly the hunters were bringing home cranial remains to extract the brains, which are rich in fats much needed for digestion by hominins with a meat-heavy diet.

In addition, the age structure of the dead mammoths does not plausibly represent random kills from a living herd of mammoths. A detailed age profile of the Spy mammoths is not available, but it is clear that roughly 74 percent of the fifty-six mammoth molars from Spy that could be aged were under the age of twelve years. In fact, at least 55 percent of these molars come from individuals under the age of two, before weaning occurs in modern African elephants. This in itself suggests a deliberate targeting of very young individuals that must have occurred and recurred for years. This is a telling characteristic. As Charles E. Kay, a wildlife ecologist at Utah

State University, remarks, "The more difficult it is for a predator to capture a particular prey, the more that predator will take substandard individuals and young. . . . So if two or more predators are preying upon the same species, the least efficient predator will tend to kill fewer prime-age animals."[21]

At Spy, bones of wolves, cave bears, and cave lions form part of the assemblage, indicating that competition among these predators and Neanderthals for food would be likely. In fact, the killing of such predators—which is implied but not proven by these bones—is a telltale sign of intraguild competition. No other comparable Neanderthal sites with large numbers of very young mammoth bones have been identified yet.

Mammoths were neither the exclusive nor even the very common prey of Neanderthals, who must have competed with modern humans for the proboscideans. Modern humans disturbed the long-standing ecosystem with their arrival about 45,000 years BP, but starting about 32,000 years ago there was a second extraordinary change. From then until about 15,000 years ago, modern humans were killing and using mammoths in extraordinary numbers not seen in any Neanderthal sites. In the Gravettian period at this time, there are many mammoth kill sites—each with between a handful and a few hundred dead mammoths. By then, Neanderthals were at best greatly reduced in population size and at worst already extinct.[22]

These mammoth megasites would benefit from redating, like many other sites in the period when Neanderthals and modern humans may have overlapped, but Gravettian sites are exclusively associated with modern humans, and a good number include modern human burials. These sites contain bones of an extraordinary number of mammoth skeletons, and some also contain processing areas where mammoths were systematically butchered. Some also have an unusually large number of wolf individuals, which bespeaks competition between this top mammalian carnivore and the modern human hunters that invaded Eurasia. One of the first things an invasive predator does is to start killing off or driving away the closest competitors that existed in the habitat before the invasion. Unlike at Spy, in Belgium—the lone mammoth megasite made by Neanderthals—Gravettian mammoth kills at these sites are not heavily concentrated on extremely young individuals.

Finally, a significant number of the mammoth megasites contain huts or shelters made of carefully stacked mammoth bones that were probably covered by mammoth hides and supported by saplings (see Figure 6.1). Were some of the mammoth bones at these sites scavenged from already-dead mammoths to be used as a building material in a nearly treeless steppeland? Quite possibly. A hut made of recently dead mammoth bones and hides would have produced an overwhelming odor that would certainly be offensive to today's living humans, but we cannot assume our forebears had the same sensibilities. Relatively few of the bones in the mammoth huts show evidence of carnivore gnawing and tooth marks, indicating either that the temperature rendered the bones unattractive or, more probably, that old, naturally cleaned bones were more often chosen as building materials. Extreme cold would also have slowed or temporarily halted the process of decay. As well as scavenging mammoth remains as building materials, there is clear evidence that modern humans killed large numbers of mammoths for their meat, hide, and fat.

Figure 6.1. These images show what a mammoth bone hut at a Gravettian site in central Europe might have looked like 30,000 years ago at Mezhirich, Ukraine. The remains of about fifty such huts have been found. *Above:* The main scaffolding of the hut is composed of tree branches, tusks, and bones of mammoth; the dwelling was probably covered in hides, but those are not shown here in order to reveal the construction of the hut. From the doorway, smoke from a hearth and a painted mammoth skull are visible. *Below:* A side view of the hut shows a zigzag pattern created by careful placement of mammoth jaws.

There are several sensitive methods for attempting to evaluate the degree of overlap in the dietary resources exploited by Neanderthals and by modern humans. Numbers and species of prey remains at sites assigned to each hominin and isotopic studies of the composition of hominin bones both indicate a very heavy reliance on a meat-based diet. Although there are indications that both hominins also used plant foods or marine resources or both, intraguild competition among Neanderthals, modern humans, and the indigenous carnivores of this time period was inevitable.

With this evidence in hand, we can ask: What would show that the invasion of modern humans created a crucial intensity of competition with Neanderthals?

7

WHAT DOES AN INVASION LOOK LIKE?

We need to stop and look at how invasions occur and what evidence of an invasion might be gleaned from the fossil and archaeological record. Once an invasive species appears, its success is predicated on its ability to reproduce and grow as a population and to take its share (or more) of the available resources. A dramatic rise in population numbers should be visible in the fossil and archaeological record. Certainly the appearance of predatory early humans in Eurasia and their eventual rise in numbers are beyond question.

The addition of a new predator is an enormous perturbation on an entire ecosystem. As William Ripple of the University of Oregon and a group of top ecologists have written in a review of the status of large carnivores in the world, "Classically, the effects of large carnivores are thought to extend down the food web to herbivores and to plants, but we are learning that their cascading influences propagate broadly to other species as mediated by their controlling effects on mesocarnivores. . . . Large carnivores have the dual role of potentially limiting both large herbivores through predation and mesocarnivores, through intraguild competition, thus structuring ecosystems along multiple food-web pathways.

Together, these controls influence the nature and strength of ecosystem functioning."[1]

As a general rule, the appearance of an invasive species is considered to be a major cause of extinction of an indigenous species. Extinctions or changes in the abundance of the prey species are obvious consequences of an invasion by a predator, because more predators will eat more herbivores. What is less obvious is that suppressing herbivore numbers in turn increases plant diversity and numbers. Finally, extinctions should also occur most commonly among those species most similar ecologically to the invasive species, their prime competitors. What often happens with a successful invasion by a predator is a top-down trophic cascade, a series of changes that reverberate through the entire ecosystem; removal of an apex predator may cause a bottom-up cascade.[2] Trophic cascades have been documented to occur in seven of the thirty-one modern ecosystems that include large carnivores today.[3]

Let's take a look at a well-documented "invasion" of a predator and the trophic cascade it provoked on a familiar ecosystem. If you have not seen Yellowstone National Park for yourself, a few numbers may help you appreciate what this example demonstrates. We tend to think of parks, even national parks, as relatively small areas. Not true. The Greater Yellowstone Ecosystem covers an area of 72,800 square kilometers, about the size of the Republic of Ireland. The park itself is 8,991 square kilometers. The ecosystem is vast and wild, and park policy now discourages any interference from humans.

Before the designation of Yellowstone as a national park in 1872, many of the indigenous tribal peoples who once inhabited the area—Shoshone, Bannock, Nez Perce, Flathead, Crow, and Cheyenne—were killed or run off in a series of brutal military campaigns known collectively as the Indian wars, though small aboriginal populations continued to live in the area and hunted in or around the park. The presence of "wild renegade Indians," as they were characterized, was seen as incompatible with white settlement, and many, largely successful, attempts were made to exterminate the tribes or confine them to reservations.

The incoming white human settlers functioned as an invasive predator and promptly eliminated their chief remaining rivals, the wolves. As humans claimed the open land for ranching or planting, a policy of getting rid of wolves was openly endorsed by the federal government. In 1915, Congress established the Federal Bureau of Biological Survey and its Division of Predator and Rodent Control, with the express mission of eliminating wolves and other large predators from all federal lands. Not surprisingly, explorers, traders, ranchers, and settlers of the American West were delighted. Gray wolves had been part of this ecosystem for millennia but were effectively extinct by the 1930s because ranchers and farmers settling the West killed these dangerous competitors preferentially.

The removal of wolves, the indigenous apex predator, made a huge difference to the ecosystem. Before the decimation of the population, wolves and the aboriginal American Indian population apparently kept elk—now about 80 percent of the herbivore fauna—in check.[4] For example, wildlife biologist Charles Kay of Utah State University cites records from twenty-six separate expeditions that traveled through the Greater Yellowstone Ecosystem between 1792 and 1872. In 369 days of travel documented by these groups, elk were sighted only twelve times.[5] They were clearly uncommon.

As soon as wolves were eliminated, the elk population soared to 19,000 individuals. They were judged "overabundant" because their browsing caused severe degradation of the rangeland. Also, where historic photographs showed tall stands of willows and aspens, in the wolf-free period these plants became heavily browsed and occupied only 5 percent of their historic range in this ecosystem. A controversial program of deliberately reducing the elk, pronghorn, and bison by shooting and trapping animals began. By the 1960s, elk populations had dropped by 75 percent to about 4,000 animals. An unanticipated consequence of lowered elk numbers was that elk hunting outside the park—once a popular pastime and tourist attraction—was virtually impossible. By 1969, the reduction program was stopped, and management efforts turned to more natural approaches. Elk numbers again increased, and hunting resumed outside the park.[6]

Coyotes, once direct competitors with wolves for prey, became so numerous that their populations swelled to the highest densities recorded anywhere. This was a direct effect of being freed from the suppression of the larger and dominant wolves.

In 1995–1996, thirty-one gray wolves from two Canadian packs were released into the park to restore the natural balance of the original ecosystem before settlement by people of mostly European ancestry (see Figure 7.1). By 2002, there were 216 wolves living within the Greater Yellowstone Ecosystem, which is judged to be close to maximum carrying capacity. Wolves have now established year-round, fiercely defended territories in the Greater Yellowstone Ecosystem (mostly within the park) despite the fact that most of the herbivores are migratory. Between 1995 and 2002, elk comprised

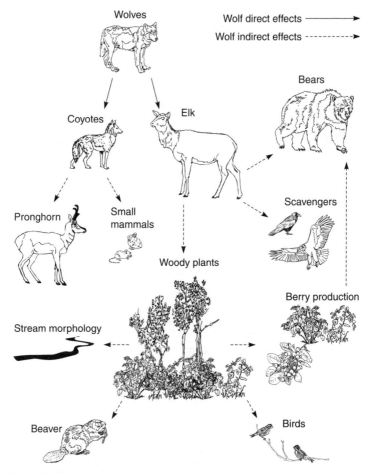

Figure 7.1. The reintroduction of an apex predator, the wolf, to Yellowstone National Park caused a trophic cascade of effects through the entire ecosystem. The changes produced by that reintroduction mimic the kinds of effects the arrival of modern humans might have had on the Eurasia ecosystem some 45,000 years ago.

92 percent of the 1,582 recorded kills. Wolves took older cow elk than did human hunters, who tended to shoot females in their prime reproductive years—if they took females at all—but both wolves and humans targeted prime bull elk. Bison were a small component of the herbivores killed. The total elk population decreased by about 7 percent in the first ten years after wolf release. Elk were also more nervous of predation and tended not to aggregate in such large herds as previously.[7]

With fewer elk, more stands of young aspen and willow survived along the watercourses of Yellowstone. These more wooded habitats favored the survival of larger numbers of birds, small mammals, and moose in the park. Absent from the park during the wolf-free period, beavers have established four colonies in the park since wolves returned. Wolves also provide winter carcasses that benefit ravens, eagles, and bears.[8] One study recorded more than 13,000 kilograms of carcasses provided by wolves in Yellowstone to other predators and scavengers on an annual basis.

Once wolves were reintroduced, elk numbers fell from an estimated 15,500 individuals to about 11,700. However, the winter of 1996–1997 was unusually severe and probably played a role in this die-off. Less than ten years later, in 2003, the elk herd numbered about 14,500.[9]

Not only prey species but coyotes—which had usurped wolves' rightful role as apex predators in the ecosystem—felt heavy pressure from the newly invasive wolves. In a study that started six years before wolves were reintroduced into Yellowstone, biologists Robert Crabtree and Jennifer Sheldon of the Yellowstone Ecological Research Center documented the abundance and hunting behaviors of coyotes in the Greater

Figure 7.2. This dramatic photograph of a wolf chasing a coyote illustrates the competition between the new apex predator and the former apex predator.

Yellowstone Ecosystem.[10] They radio-tagged 129 individuals in the northern range and followed another thirty-seven individual coyotes for a total of 200 coyote hours over five years. Coyote populations had grown to such large numbers that the eighty intensively studied coyotes of Lamar Valley were living in wolflike social packs with an average of six individuals per pack and a total of seven packs. In 1993, the Bison Peak pack in Lamar Valley reached a total of ten adults and twelve pups (due to a double litter). Average coyote density was 0.45 individuals per square kilometer on the northern range.

Almost as soon as wolves were released in 1995, they began killing and driving away coyotes (see Figure 7.2). It was as if the first item on their agenda was "get rid of coyotes," as the ranchers' and settlers' agenda had been first "get rid of Indians" and then "get rid of wolves." Quite simply, wolves would not tolerate the presence of coyote rivals in their territories—and they were equally merciless with members of other wolf packs that strayed onto their turf. In three years, coyote numbers fell from eighty individuals in twelve packs, with an average of six individuals per pack, to thirty-six individuals in nine packs averaging less than four individuals each. Coyote density dropped by up to 90 percent in core areas like the Lamar Valley, which was intensely occupied by wolf packs, and by about 50 percent overall. The depression of the coyote population improved the survival rate of fawns of pronghorn antelope considerably. Intraguild predation may account for up to 68 percent of the deaths of known cause of the predator "victims."[11] Direct interference competition is not simply theoretical but very real and very dangerous.

However, the coyotes that survived or were not attacked benefited from scavenging from elk kills by wolves in winter, as did ravens, eagles, bears, and other predators. Improved winter scavenging proved a good predictor of coyote pup survival in the spring, an important measure of reproductive success.[12]

One of the most memorable occasions of my life was when I saw the Lamar Canyon Pack of eight wolves in Yellowstone in 2012. They were feeding on a dead bison. I don't know whether they had killed it or if it had perished from mating season injuries; either is a reasonably common event. The pack ate greedily and then wandered down to the river to drink. Youngsters full of food took out their excess energy in romping, playing, and swimming. Older animals including the magnificent alpha female known as *06*—whose life history had been documented for years—snoozed in the sunshine (see Figure 7.3). We were told by a guide that, earlier in the

Figure 7.3. William Campbell of the U.S. Fish and Wildlife Service took this stunning portrait of the wolf known as *06*, who established and led the Lamar Canyon Pack in Yellowstone for several years before her untimely death.

morning, the pack had killed a strange wolf from a pack known as Mollie's Pack that lived over the hill.

Suddenly I realized that there was new activity at the carcass. I could see a wolf feeding on it though none of the Lamar Canyon Pack members had left the riverside. I alerted my companions, and we hardly breathed as we watched what happened next. Moments after I spotted the lone stranger wolf, *06* suddenly opened her eyes and twitched her ears. In seconds, she was up and moving back to the carcass, followed by her entire pack. They set out to pursue the stranger wolf that had dared to try to steal a meal from their bison carcass. The chase was drawn-out and heart-stopping—the Lamar Canyon Pack was gaining, then growing tired as the lone wolf was increasing its lead; then another, fresher, wolf was heading up the chase and drew closer to the intruder. I did not know whom I was rooting for in this race to death. The chase ended when one of the wolves caught the stranger by the tail. Other Lamar Canyon wolves came up and leaped into the fray. My view of the end of this encounter was somewhat blocked by bushes, but fur flew as wolves pounced and wrestled. From all I could tell, they pulled the stranger wolf to pieces. It certainly never stood up and walked away.

The Lamar Canyon Pack went back to snoozing in the sun near the river, youngsters a bit excited by all the action. Within an hour or so, our guide spotted a large grizzly bear coming out of a patch of forest higher on the hill. It headed straight for the carcass, sometimes walking determinedly and sometimes breaking into a lope. There was no question that the grizzly had scented the carcass.

As the grizzly lumbered closer and closer, *06* again sensed the presence of an outsider. The pack rallied to defend their carcass. They challenged the grizzly while we watched, fascinated. The wolf pack surrounded the grizzly, and their close proximity to each other made the size difference obvious. At perhaps 400 kilograms, the bear was much more massive than any wolf. (Wolves average about 50 kilograms.) The wolves darted in and nipped at the grizzly, which responded with powerful swipes of its immense paws. Very quickly, the wolves broke away from the fray, as if they had suddenly and simultaneously decided fighting was not worth the risk. They casually relinquished the carcass to the grizzly, returning to the river's edge while the grizzly went up to the carcass. This is the usual outcome of a grizzly–wolf confrontation. Grizzlies dominate all but the largest wolf packs in most encounters.

What happened next was equally interesting. First, the bear circled the dead bison and rolled on it repeatedly, feet in the air, presumably to spread its scent and establish ownership.

The bear looked downright comical. Then it took large bites of skin, meat, or bone, ripped them off the carcass, and swallowed. After only a short while, the grizzly meandered off again.

The wolves' beauty, their playful socialization, and their cozy family life contrasted sharply, even violently, with their absolutely deadly and merciless pursuit of an interloper. Their ferocious response to the strange wolf and their more measured attitude toward the grizzly were distinctly different, though the same carcass was at stake in both instances. In fact, at the end of each encounter, there was still plenty of edible food left on the carcass. I am sure the ravens, eagles, foxes, coyotes (if they dared), and other animals all got some of the leftovers. And I expect that both the grizzly bear and the Lamar Canyon Pack came back for second or third helpings later as well. Wolf 06 was an able leader of her pack, strong and clever, and she is one of the few wolves ever seen to single-handedly bring down an adult elk. She had been radio-collared, so data on her whereabouts were relayed to wolf researchers every half hour until her death on December 6, 2012, less than six months after I had seen her. Perhaps predictably, 06 was legally shot by a Wyoming hunter fifteen miles outside the park—much farther than she usually went. I and many others who had seen her mourned. She was a wolf to admire.

Witnessing these remarkable encounters helped bring home to me what the restoration of an apex predator meant and how closely it parallels an invasion by a top predator. No matter where the ecosystem is or what the staple sources of food are, adding a successful top predator to the mix will shake things up. It is worth noting that reintroducing wolves to the Greater Yellowstone Ecosystem was not a continental invasion—as these very wolves had been translocated from Canadian packs—but simply a geographic expansion of a species into a protected area in which human interference was minimal. Historically, even during the wolf-free period in Yellowstone, some Canadian wolves occasionally had ventured down into the United States, where human densities and perhaps dislike of wolves was greater. Trophic cascades can and do happen, even in geographically unbounded territories, with far-reaching effects on even a large and complex ecosystem.

We can isolate several distinct effects triggered by the invasion of wolves into the Yellowstone ecosystem. First, the wolves killed their closest competitor preferentially. After the reintroduction of wolves, every observed killing of a coyote by a wolf or wolves occurred over carcasses that had been killed by wolves.[13] Intraguild predation is indeed a major cause of death among some carnivore species.

The second obvious effect is the reduction of the density of the closest competitor in its territory. Coyotes were killed and not replaced. Coyote density dropped dramatically, by 50 percent overall and by up to 90 percent in core wolf areas in Yellowstone. Intraguild competition is often expressed as a strong negative correlation in the abundances of two competing species. Not only did the density of the competing, indigenous predator drop, its overall geographic range shrank.

On the positive side, the introduction of a new apex predator, the wolf, caused a dramatic increase in the availability of carcasses for scavenging. Because they are such able hunters, wolves in Yellowstone killed more prey during the hard winter months than they could eat, leaving more uneaten portions for scavengers like ravens, eagles, grizzlies, coyotes, and others. Elk carcasses are a major winter source of food in the ecosystem for many species, and wolves (when present) kill the majority of adult elks.[14] Wolves provided more scavenging opportunities and more meat for scavengers than human hunters did.

Whether it is the drop in population size, territory, resource share, or simply safety, a new apex predator often lowers the reproductive rate of its closest competitor. If the competitor's reproductive rate was slow anyway, this could be a major detriment.

Of course, a new predator has a negative effect on prey species. Their numbers fall, and, as shown in Yellowstone, their behavior—where they eat, when, and for how long—may change as a landscape of fear is created.

The ecologists who study this dangerous situation give us an important lead to follow in investigating Neanderthal extinction. Apex predators fill a special ecological role and are especially influential in restructuring ecosystems in times of climate change.[15] Trophic cascades affecting other species occur because, under these conditions, apex predators act as a sort of multiplier of climate change.

Does the trophic identity of modern humans as an invasive top predator operating during a period of fluctuating climate explain Neanderthal extinctions? We can outline types of data that would support either the climatic hypothesis or the human invasion hypothesis. But remember, the two hypotheses are not mutually exclusive. What I suggest here is that the synergy between climate change and the invasion of modern humans created an untenable situation for Neanderthals even though they had already survived for millennia, some periods of which saw spells as cold as the worst in MIS 3.

If competition from modern humans was a primary (but not necessarily the sole) driver of Neanderthal extinction, we can outline several types of evidence that should be found. For competition to occur, the indigenous and invasive species ought to share the same trophic level and food resources. This is already established as a fact from faunal analyses and isotopic studies. While this coincidence makes a prima facie case for intraguild competition between the two species, scientific rigor demands more detailed evidence of that competition.

One of those effects would be a diminution of Neanderthal populations after the invasion by modern humans, followed by a distinct rise in the size of modern human populations over time. Below I discuss studies designed to determine whether this prediction is fulfilled.

Another line of evidence to attest to competition between Neanderthals and modern humans would be an increase in the geographic range of humans and a decrease in the geographic range of Neanderthals. Out of the suitable areas available to them, modern humans ought to occupy an ever-greater geographic range while Neanderthals would be expected to occupy an ever-smaller percentage of the available, suitable habitat.

If modern humans were highly competitive and highly invasive top predators, we also ought to see extinctions of other predators in response to the pressure posed by this new and capable member of the guild. If we look back to when modern humans invaded the Yellowstone ecosystem, we can see it did not take long before wolves were locally extinct. Now humans in that ecosystem—particularly those who raise livestock—are concerned that the return of wolves may have a deadly effect on their way of making a living and even on their children.[16] These fears may or may not be realistic, but the deep emotions reflect the competition between wolves and humans. Now that wolves are back, humans feel their livelihood and safety are threatened and coyotes are much scarcer.

Maybe the arrival of modern humans in Eurasia made them feel frightened of or nervous about the presence of Neanderthals, who must have seemed both familiar and yet oddly different. We can only imagine how unsettling the experience of encountering another hominin species would have been—perhaps something akin to culture shock, which typically occurs when you arrive someplace where you do not speak the language, where people's clothing, behaviors, and lifestyle seem utterly inexplicable. People today are often frightened of strangers; how much more threatening would meeting another hominin species be? Limited resources and food competition would only heighten the fear. Possibly there was no conscious awareness of competition between Neanderthals and modern humans, but equally possibly, there was. At any rate, Neanderthals went extinct after the arrival of modern humans and possibly not long after.

Does this mean that climate change did not cause the Neanderthal extinction? Not really. The two hypotheses are not mutually exclusive. If climate change were the primary driver in Neanderthal extinction, we ought to see independent evidence that climate change occurred. This is abundant and already well proven by the MIS 3 data, by changes in pollen and vegetation, by the varying ranges of cold-adapted versus warm-adapted mammals, and by the ratios of ^{16}O and ^{18}O in ancient ice. Climate did change during MIS 3 and we know it.

A side effect of climate change is that the shifting ranges of climatic or habitat zones ought to parallel the shifting ranges of Neanderthals and their prey. If the climate change were so dramatic as to cause Neanderthal extinction, when the species had been successful for so long and through many other cold spells, then climate change should have been sufficient to cause a contraction of the species' range prior to extinction. It should be apparent that the available suitable habitat contracted in response to climate change, forcing Neanderthals either to adapt to a new habitat, to leave the area, or to die out. Contraction both of Neanderthal range and of their prey have been proven. Despite some uncertainties about the exact dates of various Neanderthal sites, we know that the species geographic range diminished and that the broad outlines of their diet remained stable, as expected.[17]

Another possible line of evidence that we might expect to see is increasing numbers of physical and anatomical signs of stress among Neanderthals over time. The evidence might take the form of bone and dental enamel defects—a sign of malnutrition or infection while the teeth or bones were forming. In fact, physical indicators of stress or disease in teeth and bones are fairly common among Neanderthals.[18] However, there is no way to determine whether these stress indicators were brought on by climate changes or by intraguild competition from modern humans or from other predators.

Even before the invasion of modern humans, Neanderthals may have been near the carrying capacity of the ecosystem, rendering any increase in competition more potent. We know that the Iberian site of El Sidrón, recently dated to 48,400 ± 3,200 years (calibrated), has yielded the remains of at least twelve Neanderthal individuals: three men, three adolescent boys, three women, and three infants. All of the individuals show the defects in dental enamel known as hyperplasia—resulting from starvation or disease—and some experienced several such stressful episodes while their teeth were forming.[19] Further, cut marks on the El Sidrón Neanderthal bones—including those possibly inflicted during the skinning of the crania—and impact scars from the disarticulation of long bones indicate that these bones were processed by tool-using hominins. There is no evidence of modern humans at the site, and the date does not clearly fall within the period during which modern humans were present in that immediate area. Thus there is no particular support for interpreting this site as showing intraguild violence or, necessarily, the effect of intraspecific competition. The authors conclude that the evidence of processing combined with high levels of dental hyperplasias in the sample strongly suggest survival cannibalism: that individuals ate members of their own species rather than starving to death.[20] Similar claims have been made for cannibalism at Moula-Guercy, France, about 100,000 years ago, long before modern humans reached Europe.[21]

Because being an invasive predator in a time of climate change has a multiplier effect on the invader's impact, we ought to see dramatic ripple effects of a trophic cascade among the primary consumers or herbivores in the ecosystem, analogous to what has been seen in Yellowstone. In fact, what we see in Pleistocene Eurasia is a huge crash in the predator guild after the arrival of modern humans. What the fossil record shows is the local extirpation or complete extinction of cave lions, cave hyenas, cave bears, lesser scimitar cats, leopards, and dholes—and the extinction of the indigenous hominin predator, the Neanderthal. This massive faunal change is the signature of a trophic cascade triggered by competition and invasion.

What other evidence can we use to clarify the cause(s) of Neanderthal extinction? One of the most interesting, and detectable, is the predicted change in population size and range of the indigenous and invasive competing species.

8

GOING, GOING, GONE . . .

We know that Neanderthals and moderns humans shared a trophic level as hunters and that the arrival of a new predatory species is a particularly powerful trigger for change. There is a lot of very basic evidence that both Neanderthals and modern humans were hunters of big game. They have left us their tools, the bones (sometimes modified by processing or cooking) of their prey, and, in the case of early humans at least, paintings, sculptures, and engravings of some of the animals they hunted and ate. Both hominin species can be safely assigned to the guild of predators, but more than this degree of overlap is needed. After all, wolves and mountain lions coexist in the same habitat, as do African hunting dogs, cheetahs, and hyenas in theirs. All middle to large-sized carnivores sharing a geographic range usually compete with each other to some extent.

What more can we say? In broad terms, Neanderthals and modern humans hunted many of the same prey species. This poses an odd problem in attributing Neanderthal extinction to climate change. If the habitats suitable for hunting by Neanderthals contracted, and their prey became scarce, why didn't the habitats and prey populations of modern humans also diminish? Neanderthals in particular favored wild horses and reindeer but also took aurochs and bison, red deer (elk), reindeer, woolly rhinoceros, and wild boar, as did modern humans. In a paper entitled "Eat What There Is," archaeologist Ofer Bar-Yosef of Harvard University wrote, "The contention that there was essentially no difference between Middle and Upper Palaeolithic humans, except for very rare cases, resembles our daily behaviour in the modern world. We shop for food in the nearest supermarket. With modern means of transportation, whether we live in town or a suburb, special commodities can be obtained from particular shops (and often these products represent the results of the exchange of plants and animals between the Old and the New World). Obtaining resources from distant localities seems not to have been an option in the world of Middle Palaeolithic humans. Upper Palaeolithic humans behaved in the same way although there are examples of the use of good quality lithic raw material from distances over 100 km."[1] To establish whether this localized intraguild competition was sufficient to doom Neanderthals requires additional data.

Paul Mellars and Jennifer French of Cambridge University tackled the competition question by carrying out a meta-analysis of over 200 archaeological sites in a 75,000 square-kilometer area of the Dordogne, a well-studied region of France with a high density of excavated sites.[2] (By coincidence, this region is very comparable in size to the Greater Yellowstone Ecosystem.) The Dordogne sites range in age from about 55,000 to 35,000 years ago, but the usual cautions about possible inaccuracy of radiocarbon dates calculated decades ago apply here, too. Thus, the sites roughly spanned the period of overlap between modern humans and Neanderthals in this region. Because not all sites in this region have yielded skeletal remains, Mellars and French made the pragmatic assumption that sites with stone tools of the type known as the Mousterian of Acheulian tradition and the Châtelperronian industry were made by Neanderthals. Many paleoanthropologists are willing to accept the association of the Châtelperronian with Neanderthals, but in recent years this assumption has been rigorously questioned. One problem is evidence from new radiocarbon dating that different stratigraphic layers at two of the key sites linking the Châtelperronian to Neanderthals, Arcy-sur-Cure (also known as St. Césaire) and Grotte du Renne, were mixed.[3]

Mellars and French similarly assumed that the sites yielding Aurignacian tools could be attributed to modern humans, a very common and defensible assumption. In fact, the spread of Aurignacian sites has been used by some archaeologists as direct evidence of the spread of modern humans themselves. However, the analysis by Mellars and French was criticized for the obvious reason that we cannot know whether one species and one species only made all of the sites with a particular tool type.

To see how much difference the assumption that Neanderthals created the Châtelperronian made to the outcome of their analyses, I repeated their work while eliminating information from Châtelperronian sites. My unpublished reanalysis showed that this assumption made little difference to the conclusions they reached.[4]

Mellars and French compared the number of sites made by each species, the size of the sites, the density of stone tools at each site, the numbers and identities of prey species' bones left on the sites, and the meat weight represented by the prey species (see Figure 8.1). Each of these measures is a proxy for population abundance. The objective of their investigation was

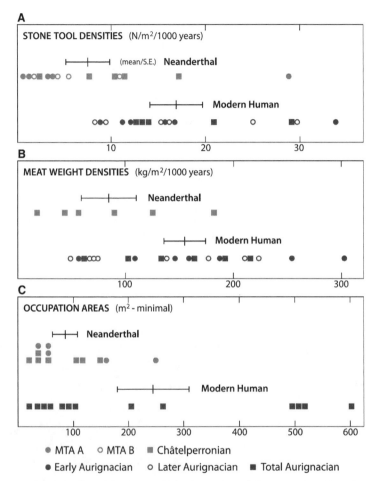

Figure 8.1. Paul Mellars and Jennifer French compared the stone tool densities (*a*), the meat weight densities as estimated from bones (*b*), and the total size of sites (*c*) occupied by Neanderthals and early modern humans between 45,000 and 35,000 years ago. These indicators of population size show that modern human populations increased sharply as soon as they arrived in Eurasia, while Neanderthal populations increased more slowly or even decreased.

to see whether there were identifiable trends over time that would indicate changes in population size of each species.

Their results were striking. There are so many more modern human sites than Neanderthal sites that an increase of about 2.3 times in the total hominin population of the focal region after the arrival of modern humans is implied. Which of the two species increased faster, as evidenced by its sites? From the older Mousterian of Acheulian tradition and Châtelperronian sites created by Neanderthals, Mellars and French documented a total increase of about 42 percent over the time span considered. Early modern human sites provide a sharp contrast. When modern humans first invaded Eurasia about 50,000 years ago, there were obviously no modern human sites. By 35,000 years ago, there were 147 such sites: a massive increase. By far the fastest increase was due to the arrival of modern humans.

More individuals need more tools. Therefore, another proxy for population size used by Mellars and French is the density of stone tools in each square meter per 1,000 years. In each thousand years, the Neanderthal sites showed a slow increase from seven to ten tools per square meter. In contrast, the density of modern human tools at Aurignacian sites rose from zero to eighteen tools per square meter per 1,000 years. These data, too, suggest that modern human populations increased perhaps 1.8 times faster during this time period than Neanderthal populations.

More individuals need more meat, too. As another indicator of relative population sizes of the two hominins, Mellars and French estimated the meat weight represented by the animal bones at the different types of sites. Because there were no suitable data for the animal bones at the Mousterian of Acheulian tradition sites, Mellars and French simply compared the meat weight of animals preserved at the Châtelperronian (Neanderthal) sites with those at the Aurignacian (modern human) sites later in time. In Châtelperronian sites, there was an average of about 85 kilograms of meat per square meter per 1,000 years. The Aurignacian sites preserved almost twice as much meat, nearly 153 kilograms per square meter per 1,000 years. These data agree fairly closely with the population estimate based on stone tool densities: there seem to have been about 1.8 times as many modern humans as Neanderthals.

Finally, Mellars and French turned to the sheer size of the sites. Clearly, a set group of people can leave a very small site or a very large site, depending on how long they occupy the area and the limitations imposed by local geographic or topographic features. Also, of course, the size of an excavated site may depend on the intensity and duration of the research, as excavations always start small and grow through time. Their compilation showed that the smallest sites produced by Neanderthals were comparable in size to those produced by modern humans. However, the largest sites produced by modern humans were about 2.4 times larger than the largest Neanderthal sites.

Because the measures they used are independent proxies for population size, Mellars and French argue that these measures ought to be added together. This procedure yields an estimate that the total hominin population increased by nine or ten times over this period. Most of this staggering population increase must be attributed to the success of modern humans. Whether or not the precise numbers they derived are correct, the bottom line of this study is obvious. All these data suggest strongly that, once early humans hit Eurasia, their populations increased rapidly and soon surpassed those of Neanderthals.

Neanderthals may have been scarce prior to the modern human invasion, an idea which is supported by many types of evidence.[5] Archaeologist Nicholas Conard of the University of Tübingen has led excavations in the Swabian region of Germany for years; his team has turned up many important finds. Conard refers to the "Population Vacuum Model."[6] He summarizes the situation: "There are many indications that late Neanderthals existed in very low population densities relative to those of [early humans] shortly after their arrival in many parts of Europe. In Swabia, for example, the densities of faunal remains, lithic artifacts and virtually all classes of anthropogenic material are much higher for Aurignacian deposits than for Middle Paleolithic [Neanderthal] find horizons. This is the case whether one considers finds per unit time or per unit sediment volume. Only rarely . . . are Middle Paleolithic deposits of comparable richness as those of the Aurignacian. Other lines of evidence suggest that Middle Paleolithic Neanderthals occupied a niche that was based on low population densities and low intensity extraction of resources relative to those of the Upper Paleolithic humans."[7] Conard and others suggest that early humans arrived in central Europe after a particularly harsh climatic phase (Heinrich Event 4), triggered at least in part by the enormous volcanic eruption that deposited the Campanian Ignimbrite ash. The ash is now dated to 39,300 years ago; both Neanderthals and modern humans were in central Europe before this eruption, but there is no firm evidence showing that

Neanderthals survived it. In fact, the youngest Neanderthal sites sampled in the recent redating of the end of the Mousterian were made between 41,030 and 39,260 years ago—just before or shortly after the Campanian Ignimbrite.[8] If western and central Europe were almost empty of other hominins, modern humans may have spread very rapidly through the area and increased hugely in population size while Neanderthal populations plummeted or withdrew to other areas.[9]

Was Eurasia essentially devoid of Neanderthals when modern humans arrived, as Clive Finlayson has also suggested on the basis of climatic data? Conard's team uncovered another key piece of evidence. In addition to recovering many tools, the Swabian excavations of modern humans levels have turned up the earliest known musical instruments—flutes made out of bird bones—and a series of exquisitely carved, small animal figurines, all of which document that a dramatic cultural change occurred. Further, his team documented the existence of a sterile layer of sediment, devoid of archaeological remains, interposed between the Neanderthal Mousterian layers and the modern human Aurignacian layers. This sterile layer is direct evidence of a period when neither hominin used the cave (see Figure 8.2).

"I was initially skeptical of the proposed stratigraphic break between the latest Middle Paleolithic [Neanderthal] and the Aurignacian [modern human layers]," Conard confessed, until fieldwork by his team at the caves of Geissenklösterle and Hohle Fels confirmed this gap in archaeological evidence. Redating of Geissenklösterle using the latest ultrafiltration methods showed that the earliest Aurignacian in the Swabian Jura was between 43,060 and 41,480 calibrated years BP. The sterile layer below the modern human occupation layer

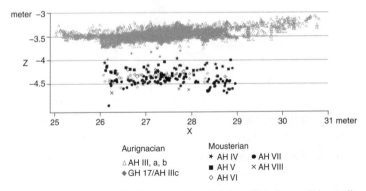

Figure 8.2. Plotting the depth of excavated finds at Geissenklösterle revealed a sterile layer, during which the cave was uninhabited, between the Mousterian (Neanderthal) and Aurignacian (modern human) layers. New dating suggests this uninhabited period may have lasted between 1,000 and 2,000 years.

lasted 2,000 or more years, when neither modern humans nor Neanderthals lived at the site.[10] There is a similar gap in various cave sites in Gibraltar between layers representing Neanderthal occupations and modern human levels.[11] If such a gap is shown to occur more generally in Europe, then modern humans may indeed have walked into a thinly populated Europe with few, if any, Neanderthals in the geographic region.

Less direct but also convincing evidence for a rapid decline in Neanderthal population numbers after about 40,000 years ago comes from a study of the genetic variability of ancient mtDNA led by Love Dalén of the Swedish Natural History Museum in Stockholm. Dalén and his colleagues sampled thirteen Neanderthals in Eurasia, testing bones from skeletons of individuals who died both before and after the modern human invasion.

Working with mtDNA offers certain advantages. Mitochondrial DNA is carried solely by the maternal line; fathers make no contribution to the mtDNA of their offspring, so the implications are easier to understand. Also, in each cell in the body, there are many more copies of mtDNA than nuclear DNA, making mtDNA easier to recover from ancient bones. Dalén's study documented a drop in Neanderthal genetic diversity in western Europe after 50,000 years ago, which the team interpreted as a partial extinction event or population bottleneck, after which a small surviving group of Neanderthals recolonized parts of Europe until modern humans arrived about 45,000 years ago. Far fewer mtDNA lineages existed after the bottleneck than before it.[12]

Within this region of southwestern Europe, then, these studies show all the attributes we would expect if there were real and potent competition between modern humans and Neanderthals. There was an extensive overlap in their prey species and a dramatic increase in the size of the modern human population compared with that of the Neanderthal population over time.

If Neanderthals were not rare in Eurasia by the time modern humans arrived, they soon became so. And yet, if intraguild competition and intraguild killing were the primary causes of Neanderthal decline, we should see evidence of the direct killing of Neanderthals by modern humans—and we don't. We do not see Neanderthal bones in archaeological sites made by modern humans, unless the Châtelperronian site of St. Césaire is a site made by modern humans. We rarely find Neanderthal bones that give even a slight indication of aggressive actions by humans. This is particularly puzzling because there is increasing evidence that, at particular times and places, Neanderthals killed and/or ate their own kind, possibly in response to starvation situations.[13] Survival cannibalism can certainly be interpreted as a response to dwindling food resources. We also know from the high frequency of dental hypoplasias in Neanderthals and evidence from Moula-Guercy and El Sidrón, discussed earlier, that Neanderthals had difficulty surviving before the appearance of modern humans in Eurasia and before the climatic instability of MIS 3.

There are two studies that do provide strong support for human–Neanderthal violence. One is a study, based on a possibly fatal wound on a Neanderthal rib from Shanidar, Iraq, by Steve Churchill of Duke University and colleagues. After studying the damage to the rib and attempting to replicate it (using a dead pig and replicated weapons), Churchill and colleagues concluded that the wound had been made by a thrown spear: an archaeological signature of modern humans.[14] Another exception comes from the modern human site of Les Rois, in France, which has yielded Aurignacian tools, many reindeer and horse bones, three hearths, the jaw of a modern human, and one of a Neanderthal child that bears cut marks made by stone tools.[15] Fernando Ramirez Rozzi and colleagues reanalyzed the bones from Les Rois and found that the cut marks on the Neanderthal may indicate it was killed and eaten, because similar cut marks were found in the same location on jaws of reindeer from the site. Obviously, reindeer were extremely likely to have been killed and eaten. However, as the analysts remark, use of the child's jaw for some symbolic purpose might have created such marks. The data are not strong enough to permit a cautious scientist to choose between the alternative explanations.

Perhaps modern humans had so many technological and physiological advantages that they had no need to kill

Neanderthals. Alternatively, Neanderthals may have avoided the territory occupied by the new apex predator, like coyotes shunning wolf-rich areas. Finally, expecting more than a handful of sites to show convincing evidence of human–Neanderthal violence may be simply unrealistic.

Clive Finlayson agrees that Neanderthal numbers dropped after modern humans arrived, but he attributes this outcome to a different cause. According to Finlayson, the reason Neanderthal population size fell was because of the colder, less hospitable climate that pushed Neanderthals to retreat into smaller, isolated areas where their ambush style of hunting worked best. Fundamentally, he is arguing for habitat loss as an extinction cause, saying Neanderthals simply went elsewhere where the weather and hunting were better. As Finlayson eloquently puts it,

> So many years of overinvestment in a body capable of handling large mammals came with a price tag. The penalty was the inability to survive where there was no cover or where long-range movements to find herds were needed. When cold pushed the tundra south and aridity pushed the steppe west, a new environment was created—the steppe-tundra. A new set of animals appeared on the scene. These animals included the woolly mammoth, the woolly rhinoceros, musk ox, reindeer and the Saiga antelope and they thrived in the expanding treeless environments that swept across Europe right down to France and northern Iberia. . . . The crucial difference was access. Whereas fallow and red deer could be stalked and ambushed, the Neanderthals would have stood out at a distance on the

> steppe-tundra. Getting close to a herd of reindeer was
> a different kettle of fish altogether. . . . The change
> from wooded to treeless . . . was rapid and Neander-
> thals had no option but retreat.[16]

Finlayson's conclusion that Neanderthals retreated to a
warmer refugium or core area as a result of the climatic down-
turns of MIS 3 relies on the chronology and geographic distribution of their later sites. Whether it was a retreat or simply
a dying off in place is unclear. And, as noted earlier, the age
of the sites supporting the existence of a late Iberian refugium
has been thrown into doubt by the team led by Tom Higham
of the Oxford University dating laboratory. That Neanderthals lived along the Mediterranean at various times cannot
be doubted; the questions are *when* and *why*.

In the Levant, where Neanderthals and modern humans
confronted climate change earlier in time, we get a different
picture of their climate preferences that conflicts with Finlayson's reconstructions. From about 100,000 to 75,000 years
ago (MIS 5 to MIS 4), Neanderthals and modern humans
alternately occupied various areas in the Middle East, presumably in response to climatic changes that altered the fauna
and flora of the region. In that region, isotopic studies by a
team led by Kris Hallin of the University of California, San
Diego, revealed that it was not the Neanderthals who disap-
peared as the climate grew colder but the modern humans.
In the sites at Qafzeh and Skhul, Hallin's team found that
modern humans dined more on wild goats and less on gazelles. Their ratios of ^{12}C to ^{13}C showed that the goats grazed
in dry grasslands where water was available only seasonally.
In contrast, the Neanderthals in the nearby caves of Amud

and Kebara, during the glacial MIS 4, ate more gazelles. The climate was cooler and there was year-round rain.

In other words, in the Levant, modern humans moved out when the climate took a downturn, becoming colder and drier, and Neanderthals moved in. This is the opposite of the pattern proposed by Finlayson in which Eurasian Neanderthals retreated to warmer climates while modern humans stayed on in the harsher habitats. We need to think in terms not simply of better or worse climate but of better or worse for which species' adaptations.

9

GUESS WHO ELSE IS COMING TO DINNER?

There is more to survival than simply hunting game effectively, which both Neanderthals and modern humans could do. Understanding how much each species needed on a daily basis to thrive and reproduce is central to measuring hunting success.

A standard measure of the metabolic needs of a species is its basal metabolic rate (BMR). The BMR is the energy a species needs just for growth and maintenance of life in a specific climate, regardless of its level of physical activity. All other things being equal, a modern human (or any other mammal) living in an extremely cold climate will have a higher BMR than one living in a more temperate climate. In fact, BMR correlates well with mean annual temperature in all kinds of mammals. Further, a larger animal will have a higher BMR than a smaller one.

The stocky, muscular build of Neanderthals meant they faced relatively high metabolic requirements at all times. The most demanding situation of all was to be a lactating Neanderthal female in a cold climate. Because Neanderthals were on average heavier, shorter, more compact, and more muscular than modern humans, their BMR was higher. For example,

Andrew Froehle of Wright State University and Steve Churchill of Duke University found that male Neanderthals on average weighed about 13 percent more than male humans, and female Neanderthals were about 12 percent heavier than modern female humans.[1] This was not the only difference. Neanderthals almost certainly were more active than modern humans, which would have produced a higher daily energy expenditure (DEE) than that of the slimmer, less muscular modern humans. Virtually every paleoanthropologist who has tried to estimate the metabolic needs of Neanderthals has agreed on this point: Neanderthals had higher metabolic needs than modern humans and a higher DEE. Having lower metabolic needs and a lower DEE gave humans a significant advantage.[2]

Froehle and Churchill calculated the energetic requirements of twenty-six Neanderthals and forty-five early modern humans for whom body mass could be estimated reliably. They plotted the site localities for those individuals onto worldwide paleoclimate maps. Although climate obviously fluctuated during the broad stages, they were able to divide the sites into cold, temperate, or tropical climates. They then calculated mean summer and winter temperatures for each site.

Deriving estimates of BMR and DEE for these data revealed some important differences between the Neanderthals and modern humans. The cold habitats inhabited by modern humans were on average −2.2 °C, whereas those inhabited by Neanderthals averaged 6 °C, substantially warmer. In other words, during the colder climatic regimes, Neanderthals lived only at the warmer sites. In contrast, modern humans could and did survive in harsher climates. This difference implies that the problem with climate change and competition between Neanderthals and modern humans was most acute during the colder times, an idea reinforced by the fact that there was no meaningful difference in the mean annual temperatures of the temperate or tropical areas where the two species lived. Thus, perhaps the pattern of occupation in Europe was unlike that seen in the Levant because the "cold" phases in the Levant were not cold enough to cause undue stress.

In all locations regardless of climate, the energetic requirements of male and female Neanderthals were higher than that of modern humans by about 7 to 9 percent, simply because Neanderthals were larger. Living in colder habitats imposed a greater energetic burden on both modern humans and Neanderthals than living in temperate or tropical regions. In both species, males needed an extra 1,200 kilocalories (kcal) daily in cold climates compared with warmer ones; females who were neither pregnant nor lactating needed an extra 800 kcal daily in cold climates. Children of each species, being smaller, would have lost body heat more rapidly and been more vulnerable still. Because Neanderthals had bigger bodies, their extra need amounted to an average of 275 kcal each day over that required by modern humans. However, this disadvantage was in part offset by the fact that the bigger body and greater muscle mass also offered Neanderthals a way to slow heat loss.[3]

A tremendous caloric intake would have been needed in cold climates to sustain pregnancy and lactation, the latter being probably the most energetically demanding period in a female's life. In various mammals, the extra costs of pregnancy range from 20 to 30 percent more than the normal intake of nonreproductive individuals, whereas lactation calls for an extra 35 to 145 percent over normal needs.[4] Again, because Neanderthals were heavier, their metabolic needs during pregnancy and lactation would have been greater than those of modern humans in the same climatic conditions.

Brian Hockett of the Bureau of Land Management in Nevada estimated the necessary intake of a pregnant and then of a lactating female Neanderthal in a manner that really brings the differences to the fore. He translated the necessary number of kcal (5,500 daily) into familiar terms for humans living today: "From the perspective of a modern fast food diet, a pregnant Neanderthal woman would need to eat ten large cheese burgers per day (or three in the morning, three at mid-day, and four in the evening), or seventeen orders of chicken nuggets per day (or five orders in the morning, six at mid-day, and another six in the evening). This perspective assists in understanding the amount of food that these models suggest pregnant Neanderthal women consumed on a daily basis, although it certainly does not negate that possibility."[5]

Aside from the sheer difficulty of obtaining so much protein and fat-rich food on a daily basis when McDonald's fast food was unavailable, there are other problems. With a high level of daily physical activity, what would happen if a pregnant Neanderthal woman consumed 5,500 kcal from terrestrial mammal meat every day? Hockett asserts the result would be "Dead Neanderthals. Dead Neanderthals result from this diet because of two main reasons: Neanderthals could not have consumed such large quantities of calories per day without consuming very large quantities of terrestrial animal muscle and internal organs, culminating with the end result of rather severe over- and underconsumption of essential nutrients; and . . . within the 'terrestrial mammal' component of a diverse diet based on animal types . . . there is little diversity in essential nutrient composition; this is another way of saying that it does not matter much whether Neanderthals ate multiple species of terrestrial herbivores (e.g., bison, deer, rabbit, wild goat), the only way for them to have consumed a greater diversity of essential nutrients was to consume a greater diversity of food types."[6] Hockett reminds us that leafy green vegetables and other plants must have formed a significant part of Neanderthal diets, although plants are nearly invisible in the archaeological record of Neanderthals and early modern humans. This is strictly a problem of preservation in most environments and leaves a significant hole in our understanding of hominin diets during this time.

If robust Neanderthals could not survive on the all-meat diet, how did puny modern humans manage to survive in colder habitats where vegetable foods would have been hard to find in winter? Other than dietary changes—a lot of side salads, thank you—there might have been some behavioral adaptations that would lower this prodigious metabolic need for calories.

Froehle and Churchill cite archaeological evidence showing that, starting about 45,000 years ago, modern humans had bone needles and Neanderthals did not. These needles were probably used to produce clothing from skins and hides that offered more protection against the cold than the looser, untailored skins Neanderthals may have worn. There is also some evidence that modern humans had more efficient hearths and shelters.[7] To some extent, modern humans shielded themselves from the high energetic demands posed by the cold by cultural buffering that surpassed what Neanderthals used.

John Shea has similarly proposed that the long-distance weapons created by modern humans made hunting less strenuous and more efficient, adding another energetic saving. If daily energetic requirements for obtaining food and keeping warm were lower among modern humans than among Neanderthals, more energy was left over to be expended on reproduction and child rearing, which might have fueled the population increase we know occurred. Nonetheless, staying alive was a formidable task, especially during the colder periods.

There is another important ecological factor that posed difficulties, too. The problem for Neanderthals and modern humans was not simply the existence of the other hominin, nor was it simply climate change. An additional and very serious problem was the existence of a large carnivore guild in Eurasia during this part of the Pleistocene era.

A way to understand Neanderthals and modern humans on the landscape is to treat them as part of the predatory guild. In a brilliant piece of work, Chris Carbone of the Zoological Society in London and John Gittelman of the University of Virginia showed that the density of a given carnivore species on the landscape is a function of both its own mass (body weight) and the mass of the prey species in the area.[8] Thus, a predator the size of an average male Neanderthal would need 10,801 kilograms of herbivores to support him, whereas an average male human would need only 9,412 kilograms of herbivore. These numbers are fairly similar but not identical, in keeping with a 13 to 15 percent difference in estimated body weights. On the basis of body size alone, each hominin species would live on average at a density of about one individual per three square kilometers, depending on prey availability. This predicted density is comparable to the low end of densities recorded for wolves, which are a little smaller in body size. In reality, what this figure probably means is that a small group of either hominin—say ten individuals—needed to range over a large area of about thirty square kilometers. It takes a lot of land full of prey to support such hunters.

Though predators often dictate the density of herbivores in a given area—by killing them—the opposite is also true: the number of herbivores sets a cap on the number of carnivores that can live in a region. Of course, adding an additional predator of fairly large body size, like a modern human, would produce repercussions that would ripple through all the other predators in the area and their prey.

How many Neanderthals or modern humans could live on the entire 75,000 square kilometers of western France studied by Mellars and French? The answer is that, if the sex ratio were fifty-fifty, this enormous area could house and feed only about 3,400 Neanderthals or 3,800 modern humans. Three to four thousand individuals is about three times the standard minimum viable population for modern mammals with medium to large bodies. The same area would, theoretically, support almost 5,000 wolves, because they are smaller than either hominin. These figures suggest that the invasion of a new, top predator like modern humans might have exerted a serious competitive pressure on other members of the predatory guild.

Did Neanderthal populations diminish or move out to avoid dominant modern humans or because there wasn't enough food locally for both species? Or had Neanderthal populations already diminished because of changing climate and habitats, leaving a very thinly inhabited area for modern humans to invade? In either case, intraguild competition both between these hominins and among these hominins and the other predators in the area would have been serious.

The carnivore competitors for Neanderthals and modern humans during the late Pleistocene were formidable.[9] Although there were a number of archaic carnivores that co-evolved in the Eurasian ecosystem with Neanderthals, by about 50,000 years ago the pertinent predator guild included only eight nonhominin carnivores capable of hunting medium- to large-sized prey.

Three attributes of any carnivore do much to determine its place in an ecosystem and where the competitive conflicts lie. The first is size, as body mass is an excellent predictor of the preferred prey size that a species hunts.[10] The second attribute is hunting style. Carnivores tend to be either ambush hunters, which sneak up on prey, or running predators, known as coursers, which rely on swift pursuit. Finally, another key issue is whether the carnivores hunt alone or in social groups.

A quick consideration of the predator guild in Eurasian between 50,000 and 25,000 years ago demonstrates the differences. In order of body size, the largest Eurasian predators at this time were the cave bear, *Ursus spelaeus,* which was later replaced by *Ursus ingressus* and the brown bear, *Ursus arctos.* These were large bears that existed both during and after the period in which Neanderthals went extinct. They may have posed little competition to other predators because most bears are omnivorous and eat a wide range of plant, aquatic, and mammalian foods. Bears could and apparently did minimize their dietary competition with either hominin by shifting to more omnivorous diets or by switching to different prey.

Isotope analyses show that cave bears did exactly this where they overlapped with modern humans.[11]

The next largest predator was the cave lion, which was 25 percent bigger than today's lion.[12] The cave lion was probably a formidable ambush predator with powerful limbs and claws. Scholars argue whether this species is rightly called *Panthera leo* like the modern lion, or *Panthera leo spelaea*, indicating it was a distinct subspecies, or simply *Panthera spelaea*. Whatever it is called, the cave lion was strong, large, and able. When cave lions are depicted in cave art—which is not often—they are not shown as having manes. R. Dale Guthrie of the University of Alaska at Fairbanks has made the interesting suggestion that the lack of this feature, which distinguishes modern male lions from females, may indicate that cave lions lived not in large prides but mostly as pairs.[13] However, mane size is not well correlated with pride size, so this interpretation is not secure. On size alone, cave lions would have been dominant over other felids in the predatory guild at the time and quite possibly over the canids, unless the canids were in large packs. The most probable prey of cave lions were wild horses, larger deer, bison, and aurochs.[14] This prey preference would place them in direct competition with Neanderthals, with a heavy overlap in preferred prey species.

Another felid was the lesser scimitar cat *(Homotherium latidens)*, a saber-toothed cat that was about the size of a modern tiger. Judging from the extreme paucity of fossils, the lesser scimitar was very uncommon if not effectively extinct by the time modern humans arrived. An analysis by William Anyonge of Xavier University of the anatomy of the lesser scimitar cat shows it had elongated canine teeth for which it is named—teeth very suitable for killing large, thick-skinned prey.[15] The sharp slicing cheek teeth enabled a *Homotherium* to strip flesh neatly and quickly from a carcass. Overall its dental adaptations suggest a hypercarnivorous species that ate little bone and could not crush bones for marrow. The lesser scimitar cat's legs were adapted for fast running, suggesting a cursorial adaptation somewhat similar to a very large cheetah. Anyonge suggests that this felid specialized in taking down juveniles of very large species like mammoths, large bovids, or elk.

Another large cat was the leopard, *Panthera pardus*, which is probably the same species that is alive today. Body size estimates suggest that the ancient leopard was about as large as the modern one.[16] Like modern leopards, the ancient leopards of this time were probably largely solitary and specialized in ambush hunting, preferring wooded or forested habitats. Modern leopards focus on prey species that weigh about forty kilograms,[17] which suggests the similar-sized ancient leopards probably preferred animals like smaller ibex, reindeer, or roe deer.

Slightly smaller than the leopard was the cave hyena, which is sometimes considered to be the same species as the modern hyena, *Crocuta crocuta*. Some scholars, emphasizing that its body size is 10 to 15 percent bigger than that of modern spotted hyenas, feel it deserves its own subspecies, *Crocuta crocuta spelaea*.[18] Like modern spotted hyenas, the cave hyena was robustly built with large, bone-crushing cheek teeth powered by strong jaws and muscles and had a distinct tendency to hunt in social packs in open country. Modern spotted hyenas are superb cursorial pack hunters. Because of their strength, large body size, and social nature, spotted hyenas (and presumably cave hyenas) are both active hunters and confrontational scavengers that often win in direct competition for a carcass. Cave hyenas were the only specialized bone crushers in the Eurasian guild between 50,000 and 25,000 years ago.

The final two large mammalian carnivores in the Eurasia ecosystem at that time were wolves *(Canis lupus)* and dholes *(Canis alpinus)*. Both are pack-hunting canid species found today in the Americas and Eurasia (wolves) and Asia (dholes). The wolves were roughly as large as big Alaskan wolves today and were certainly a coursing, group-hunting species. Almost any large mammal in the ancient ecosystem would have been fair game for a pack of wolves. Wolves are not well equipped with strong jaws and bone-crushing teeth, so any scavenging they do consists of taking meat, fat, hide, or small bones. Dholes were much smaller than wolves but larger than most modern dholes, and are fairly rare in the fossil record during the Middle and Upper Pleistocene. Because modern dholes hunt in packs of up to ten individuals, they can bring down surprisingly large prey up to about 713 kilograms. Hunting singly, an ancient dhole's prey would consist of small mammals, but hunting in a pack, dholes would have targeted horse, bison, aurochs, and large elk. They do not have bone-crushing teeth and so are hypercarnivorous to protect their teeth. In modern Asia, dholes frequent wooded or forested habitats, but it is not clear whether that is a recent effect of modern settlement patterns.

All of these medium to large members of the predator guild would have competed for prey with Neanderthals and modern humans. There were also several smaller predatory species, including red foxes, Alpine foxes, weasels, and wolverines, and many predatory birds, such as various eagles, hawks, crows, and vultures.

The relationship between predator body mass and prey size is constant for mammals because of the energetic constraints of hunting. Carbone and his colleagues expressed this simply, explaining: "Carnivores weighing 21.5 kg or less feed mostly on prey that is 45% or less of their own mass, whereas carnivores above this feed mostly on prey that is greater than 45% of their own mass. . . . This classification holds for 92.1% of 139 species (92.9% of the smaller mass class, 82.6% of the larger mass class)."[19] Three-quarters of the smaller carnivores are omnivorous and take insects and/or plant resources as well as the meat of vertebrates. Just over half of the larger carnivores feed purely on vertebrate prey, with most exceptions being bears or their close kin. As a general rule, a fifty-kilogram mammalian predator will focus on prey that weighs about fifty-nine kilograms.

There are significant differences in hunting style that come into play as well. Lone or solo hunters take smaller prey than do pack hunters. Ambush predators, like most of the big cats, take smaller animals than coursers, like wolves. To complicate matters further, many coursers are pack hunters, too.

Christine Hertler and Rebekkah Volmer of Frankfurt University developed equations for using the predator's body mass to calculate the size of prey that it targets, or its prey focal mass. Another equation allows the prey focal mass to be modified to accommodate differences in hunting style—solo or pack hunting—and the size of the pack.[20] Below I use estimates of the body mass (in kilograms) of the medium- to large-sized members of the predator guild in Eurasia from

Table 9.1 Mass of species in the Paleolithic predatory guild

Species	Body mass (kg)	Prey mass (solo)	Prey mass (pack = 6)
Cave bear	500	N/A	N/A
Brown bear	400	N/A	N/A
Cave lion	160–325	327–1,274	10,162–39,536
Lesser scimitar	150–230	289–656	6,331–14,366
Cave hyena	65–90	58–109	1,807–3,372
Neanderthal (ambush)	62–85	53–97	1,651–3,023
Hss (courser)	59–81	41–75	1,269–2,330
Leopard	45–90	28–108	N/A
Wolf	45–80	28–51	839–2,691
Dhole	20–40	5–19	160–602

The preferred prey size of a predator is a function of its own body mass and its hunting style. Note that a pack of six cave lions or six lesser scimitar cats require larger prey than any species in the ecosystem (see Table 9.2). This suggests they did not hunt in large packs. Larger cave lions and lesser scimitar cats would have to hunt in pairs (cave lions) or trios (lesser scimitar cats) to obtain enough food from killing a single prey in this ecosystem. *Hss* = *Homo sapiens sapiens*, or early modern humans; N/A = not applicable because of dietary or hunting habits. Data on body size come from the following sources: W. Anyonge, "Body Mass in Large Extant and Extinct Carnivores," *Journal of Zoology London* 231 (1993): 339–350; A. Turner and M. Antón, *The Big Cats and Their Fossil Relatives: An Illustrated Guide to Their Evolution and Natural History* (New York: Columbia University Press, 1997); and S. E. Churchill, *Thin on the Ground* (New York: Basic Books, 2014). Predictive equations on preferred prey size are from C. Hertler and R. Volmer, "Assessing Prey Competition in Fossil Carnivore Communities—A Scenario for Prey Competition and Its Evolutionary Consequences for Tigers in Pleistocene Java," *Palaeogeography, Palaeoclimatology, Palaeoecology* 257 (2008): 67–80.

about 50,000 to 25,000 years ago to calculate the size of prey on which they would feed predominantly, using Hertler and Volmer's equations. The pack-hunting calculations are based on a pack size of six, which is within the recorded ranges of the modern equivalents of these species.

Table 9.1 shows that, as a rule, pack hunters take larger prey than lone hunters with the same body mass. Thus, by hunting in packs, a predator can bring down much larger animals. However, the individual's share of the kill is not necessarily larger than if it were hunting alone. A dominant individual often eats more than its "fair share" of a carcass, and a subordinate will get much less. The predicted prey size of a pack of six cave lions or six scimitar cats actually exceeds the size of the largest animals in the ecosystem, so they cannot have hunted in such large groups.

If you are a lone predator, you are better off being an ambush hunter than a courser of the same size. This difference reflects the fact that ambush predators have evolved to hunt singly, by stalking close to a target and then making a rapid run or pounce. Sprint speed is essential. A lone courser uses swift pursuit, more like a distance runner, but the chase goes on for longer and expends much more energy. When fatigue sets in, a lone courser does not have the option of swapping the lead with another member of the same pack. One individual can have great difficulty in running down a lone prey.

In this ancient ecosystem, theoretically only pack-hunting lions and lesser scimitar cats could take down the largest adult mammals (mammoths, estimated at about 5,500 kilograms). However, wolves, cave hyenas, and both hominins—if hunting in packs—could take any other mammal in the ecosystem. Leopard, by analogy to their modern counterparts, were very unlikely to have hunted in packs and would rarely be able to capture an adult mammoth. But it is important to remember that all prey species start as youngsters, and catching a baby mammoth, for example, is a very different proposition than taking an adult.

What was on the menu for these carnivores? There were many prey species to choose from in the Pleistocene habitats. Table 9.2 shows body weight estimates for the more common prey species of medium to large prey in Eurasia between

Table 9.2 Size of herbivorous prey species

Species	Body Mass (kg)	Habitat
Mammoth *(Mammuthus primigenius)*	5,500	Open steppe, cold
Woolly rhino *(Coelodonta antiquatus)*	2,668	Open steppe, cold
Wisent *(Bison bonasus)*	891	Forest
Giant deer *(Megalocerus giganteus)*	670	Open woodland, steppe
Steppe bison *(Bison priscus)*	529	Dry steppe/prairie
Elk/red deer *(Cervus elephas)*	500	Woodland
Horse *(Equus hydruntus)*	335	Steppe
Musk ox *(Ovibos moschatus)*	285	Open steppe, cold
Aurochs *(Bos primigenius)*	269	Open steppe, cold
Reindeer/caribou *(Rangifer tarandus)*	100	Steppe, cold
Wild boar *(Sus scrofa)*	90	Woodland
Ibex *(Capra ibex)*	40	Woodland, alpine
Chamois *(Rupicapra rupricapra)*	40	Forest, alpine
Roe deer *(Capreolus capreolus)*	25	Closed woodland/forest

Data are from D. Brook and D. Bowman, "The Uncertain Blitzkrieg of Pleistocene Megafauna," *Journal of Biogeography* 31 (2004): 517–523.

50,000 and 25,000 years ago, arranged from largest to smallest. We know from the faunal remains at archaeological sites, where many bones are cut marked or broken to remove marrow, that Neanderthals routinely took animals much larger than would be predicted for a carnivore of comparable body size. So did modern humans.

Does that mean that there is something wrong with the equations used to predict prey size in Table 9.1? Or are the prey sizes in Table 9.2 incorrect? No, both sets of data look very reasonable and are based on comparisons to living species' anatomy. Then why do the hard facts—the bones themselves—show a much wider range of prey than one would expect hominins to take, given their size? The problem is that the calculations used to create Table 9.1 make no allowance for the ability of humans to use weapons, which apparently gave them substantial advantages over "average" pack-hunting or ambush predators of the time. Working in packs, with weapons, both Neanderthals and modern humans were able to take down mammoths and did so as well as any other animal in the ecosystem.

Finlayson has diagnosed Neanderthals as being primarily ambush hunters for good reasons. They show no anatomical adaptations for either speed or long-distance running, and their weapons were handheld, not projectile. They also seem to have preferred more closed habitats or ecotones between forest and open tundra. If his interpretation is correct, these ambush adaptations may have prevented Neanderthals from taking adult mammoths regularly, though we know they preyed on juveniles and newborns at the Belgian site of Spy.

Making a successful kill is not the end of the story. In the ecosystem of that time, either Neanderthals or modern humans would have been challenged for possession of the carcass by other predators. Confrontational competition can be ferocious and, in modern ecosystems, frequently ends in death of one of the participants. Could either of these hominins dominate one of the contemporary carnivores in a confrontational encounter? Sheer body size, which often predicts dominance, does not favor the hominins. If alone, either hominin was seriously outweighed by cave lions and lesser scimitar cats; cave hyenas, wolves, and leopard approximately equaled these hominins in body size. Lone hominins must often have been forced to abandon a carcass to any of the other predators, with the possible exception of a lone dhole.

Neanderthal expert Steve Churchill has thought a lot about Neanderthal hunting. He thinks these hominins were unlikely to have been dominant in the predatory guild, citing their intermediate body weight: "Where did Neandertals fall in the dominance hierarchy of large-bodied carnivores? This can be difficult to know with any certainty, but I think a case can be made that wherever they fell, they *weren't* the socially dominant members of their guilds." He continues, "I suspect that Neandertals may have taken on lions, scimitar cats and hyenas—either to usurp a carcass or to defend one from theft—when numbers were on their side. Given the pride and pack hunting of these carnivores, I also suspect that most times the Neandertals were on the losing side of these interactions."[21]

Working in packs and possessing long-distance projectile weapons could perhaps have enabled modern humans to dominate even these large predators on occasion, depending on relative pack sizes. But Neanderthals, who lacked long-distance weapons, would probably have been foolhardy to attempt to dominate cave hyenas or cave lions. Even a pack of Neanderthals would more often have been intimidated by the other predators because they had handheld weapons. Getting close enough to use those weapons would have placed Neanderthals within range of carnivores' claws or teeth. Neanderthals' primary strategy in direct confrontation with a predator was probably group display—gestures like hollering and waving their arms to make themselves appear bigger, perhaps throwing stones or other items—and, if possible, removing as much meat as quickly as possible and then abandoning the carcass.

The importance of weaponry to either hominin species has to do not only with killing efficiency but with processing carcasses rapidly to remove meat, fat, and marrow before any other predator arrived at the scene to challenge them. Stone tools enabled either hominin to crack bones and extract nutritious and fatty marrow that was inaccessible to most of the other predators except cave hyenas. With the very largest prey—mammoths and woolly rhinos—not even an enormous hyena could break open the marrow bones. Experiments show that modern humans armed with large stones can break open marrow-bearing bones of modern African elephants but not without considerable time and effort.[22] No other carnivore we know of can do this, though large carnivores will chew on the ends of bones to extract fat.

Though modern humans may have had the edge, they overlapped completely with Neanderthals in terms of the prey they took. For example, Don Grayson of the University of Washington and Françoise Delpech of the Centre National de la Recherche Scientifique in Talence have carried out painstaking faunal analyses of numerous archaeological sites in Europe from the Mousterian (Neanderthal) through Aurignacian (modern human) times and have been unable to show consistent significant differences in prey choice between the two species. They conclude, "The archaeological analyses document that Neanderthals and early modern humans hunted, and then processed, identical kinds of animals in extremely similar ways. Although the proportions of particular species in their diets differed significantly, this had more to do with climatically driven changes in the abundances of those animals on the landscape than with the choices made by the hunters. . . . The null hypothesis—that Neanderthal and early modern human diets did not differ in any (detectably) important ways, at least as regards the large mammal component of that diet—now has remarkably strong support."[23]

Not only did both Neanderthals and modern humans overlap with each other in their preferred size of prey, there is a tremendous amount of overlap within the carnivore guild as a whole (see Figure 9.1). For example, the cave lion, if hunting solo, would have focused on prey such as aurochs, other bovids, giant deer, elk (red deer), and horses. But of course they could have taken smaller species, too, or juveniles of larger ones (rhinos, mammoths), and probably often did. No other carnivore could take such large prey if hunting alone.

The lesser scimitar cat overlapped heavily with cave lions in terms of the lower end of the lion's preferred prey spectrum, roughly 300 to 1,300 kilograms. Leopards, cave hyenas, wolves, Neanderthals, and modern humans, if hunting solo, all overlapped with each other in their preferred prey species, which ranged from about 30 to 110 kilograms. Lone dholes would have taken smaller species still, if they were not effectively extinct by the time modern humans arrived.

Excepting leopards, bears, and possibly lesser scimitar cats, all of the other members of the predatory guild in Eurasia at this time probably hunted in groups. Pack hunting is a game changer. Working as a pack, any species can take down a larger animal than it can hunting solo. However, hunting as a pack also means the prey must be larger to feed the entire pack.

In groups, any of these predators could have taken down any of the extant prey species. The possible exception to this rule would be Neanderthals and modern humans, which—if

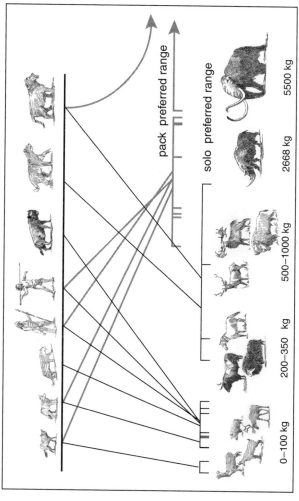

pack preferred range

solo preferred range

0–100 kg   200–350 kg   500–1000 kg   2668 kg   5500 kg

Figure 9.1. An animal's body size strongly influences the size of its preferred prey. Prey and predatory species are arranged by adult body weight. All of these predators probably hunted in packs, with the exception of the leopard. Pack hunting meant the predators were able to and needed to take larger prey. The huge overlap in the preferred prey sizes of different species suggests intense competition whether a predator hunted singly or in packs.

they had been equipped like standard carnivores—would have had difficulty with full-grown mammoths. However, if there is one thing we know, it is that neither modern humans nor Neanderthals were equipped like standard carnivores. The obvious conclusion is that there was significant and important competition within the predator guild and that all of the larger carnivores targeted the same species. The obvious exceptions are cave bears and brown bears, which are so herbivorous—according to compelling isotopic studies by Hervé Bocherens and his team[24]—that I have not attempted to calculate their favored prey size.

As Churchill points out, there are some behaviors that may lessen competition between species.[25] One species may differentiate from a competitor in terms of habitat preferences, such as using woodland areas to hunt for food versus open steppe or tundra. Isotopic analysis of cave lions, for example, indicates that their diet did not overlap with cave hyenas from the same site. Finlayson interprets the data to suggest that Neanderthals were woodland or ecotonal ambush hunters rather than open steppe specialists. If so, Neanderthals still would have competed with leopards, cave hyenas, and to a considerable extent dholes and wolves.

Another strategy for lessening competition is to differentiate from other species in terms of hunting time: preferring night hunting to dawn and dusk or diurnal hunting to nocturnal hunting. Such a shift has been documented in Nepal, where the human and tiger densities in and around Chitwan National Park are surprisingly high. In apparent response to the noise and disturbance associated with human activities such as wood gathering, tigers in Chitwan are rarely active during the day, much less so than tigers studied in similar areas in Malaysia and Indonesia. This enables the Chitwan tigers to coexist more successfully with humans.[26]

Mammalian predators likely had better night vision than either modern humans or Neanderthals; both were probably diurnal or daytime hunters. Whether this spared the hominins from much intraguild competition is debatable, as modern lions, hyenas, wolves, and dholes all hunt day and night.

A third approach to avoiding competition is to specialize in the ability to process various types of food, such as flesh-eating specialists with slicing teeth versus bone breakers with large, crushing teeth to extract marrow. Hominins specialized at using stone tools to cut off portions of meat or fat from carcasses for removal from dangerous proximity to a fresh carcass; at this, they probably excelled. Another specialization used by hominins, with the help of stone tools, was breaking bones to extract nutritious and fatty marrow. Only cave hyenas would have given modern humans much competition at bone breaking; the other predators at this time could handle only much smaller bones. Though none of these strategies were likely to have been entirely effective at shielding hominins from resource competition, they may have helped.

We can actually see the effects of the intensity of intraguild competition between 50,000 and 25,000 years ago. There was an ecosystem-wide crash involving many members of the predatory guild in Eurasia after modern humans arrived. Neanderthals, cave hyenas, and possibly dholes went extinct at about 40,000 years ago.[27] The cave lion began to decline in genetic diversity at 45,000 years ago, was uncommon or rare in Europe between 40,000 and 35,000 years ago, and then repopulated Europe to go regionally extinct at about 25,000 years ago. Two species of cave bear went extinct at about 28,000 and 25,000 years ago. The most recent lesser scimitar cat is dated to 28,000 years ago, but there are only two specimens more recent than 400,000 years ago, suggesting either that this species was extremely rare or that the date needs confirmation. Leopards are shown in the paintings at Chauvet Cave, dated to perhaps 40,000 to 38,000 years ago, but according to fossils, the decline of this species in Europe apparently began at just more than 40,000 years ago.[28] In other words, within a few millennia of the arrival of modern humans, the predator guild experienced an enormous disruption that led to the extinction or extirpation of many species.

It is impossible to be more precise about the timing of each extinction, both because of problems with dating and because we can look at the dates only of the individuals that we find as fossils; those specimens that have yet to be found are for the time being invisible. But we know that all these species competed to greater or lesser extent for the same prey species, and the invasion by another highly successful predator can, I think, explain much about these predator guild extinctions.

But what about bears? They were technically predators, but the isotope studies indicate they were omnivorous or even entirely plant eating. Thus, the evidence we have is that cave bears were not competing directly with either modern humans or Neanderthals for food. I have included these in the Pleistocene predatory guild because their extinction confirms the tremendous importance of competition as an evolutionary force.

10

BEARING UP UNDER COMPETITION PRESSURE

The habits of cave bears at this time tell a fascinating story. There were two species of cave bear in Eurasia, the eastern, *Ursus ingressus,* and the western, *Ursus spelaeus.* They can be distinguished on anatomical details and on mtDNA differences. Their extinction is a two-episode story that sheds light on the wave of predator extinctions in Eurasia at this time.

There may have been a short period during which the two cave bear species overlapped in Germany, but in what is now Austria the two species coexisted for possibly as much as 15,000 years. How did they avoid competition with each other? Both species of cave bear were entirely herbivorous, judging from isotopic studies by Hervé Bocherens and colleagues.[1] And yet, the specimens they analyzed did not overlap in their isotopic values, implying that the two species partitioned the herbivorous bear niche either by eating different plants or by eating plants derived from different regions or altitudes. In other words, though both were wholly or predominantly plant-eating cave bears, they did not compete for food and could coexist.

Cave bears were not the only ursids in Eurasia at the time; the brown bear, *Ursus arctos,* lived in the same region, complicating the ecological situation still further. The brown bears were smaller in body size but clearly much more carnivorous than either cave bear. They were not as hypercarnivorous as cave lions in the same valley. Brown bears, like their modern representatives, showed a typically bearlike flexibility in diet.

Several scholars have suggested that modern humans may have decimated the western cave bears by hunting them as they hibernated in caves, making them a predictable resource. As Susanne Münzel, a University of Tübingen expert on cave bears, sums up the situation: "[Some favor] the idea of a competition for shelter since both modern humans and Neanderthals would have been strong competitors for these caves and might have forced cave bears into less suitable hibernation dens. However, hunting played an additional role and might have had considerable impact, not necessarily on a single genetic group, but on the cave bear population as a whole. . . . For cave bears it was fatal to prefer caves for hibernation. In these dens they were easy to locate in contrast to brown bears which at least nowadays favour dens in the open landscape. . . . Furthermore, evidence for hunting of cave bears in the [region] is given by numerous cut and impact marks on cave bear bones . . . and in a thoracic vertebra from Hohle Fels with a projectile point still sticking in the bone" (see Figures 10.1 and 10.2).[2]

With tangible evidence like this, there can be no doubt that humans hunted cave bears. But the bones also tell a story of opportunism on the part of modern humans. During Neanderthal times, most of the bones of cave bears recovered from caves in the Ach Valley of Germany were from animals that died natural deaths, presumably during hibernation or early spring. Signs that cave bears were systematically hunted by Neanderthals are rare. In the same area during Aurignacian and Gravettian times, after modern humans arrived, cave bears become less common in bone assemblages, but their bones more often show marks from cutting, skinning, or defleshing with stone tools. Systematic hunting and processing of cave bears by humans had begun. Taking bears during their hibernation phase or in early spring, when newly awakened and hungry adults left the cubs to go feed, would have been a relatively easy prospect. Because both Neanderthals and modern humans used caves as shelters, knowing where caves were and which ones were suitable for habitation would have been an important piece of information for survival.

We also know that cave bears underwent a dramatic diminution in population size. The representation of both cave bears species shrank from about 55 percent of the fauna remains in Neanderthal times—most of which were probably due to natural deaths unrelated to hominin activities—to between 20 and 32 percent in modern human times (see Figure 10.3). This conclusion is supported by findings from an international research team, who extracted and studied fifty-nine mtDNA sequences from cave bears and forty from the brown bear. According to Aurora Grandal-D'Anglade of the University of Coruña, an author of the study, "The decline in the genetic diversity of the cave bear (*Ursus spelaeus*) began around 50,000 years ago, much earlier than previously suggested, at a time when no major climate change was taking place, but which does coincide with the start of human expansion."[3] Because the decline in the genetic diversity of cave bear populations started during a period of climatic stability, the team dismisses climate fluctuation as a causal agent and

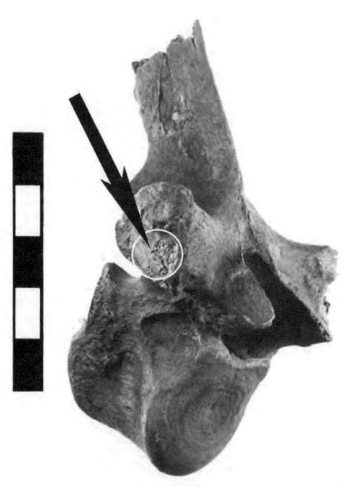

Figure 10.1. A flint projectile point embedded in this cave bear vertebra from the Gravettian (modern human) level of Hohle Fels, Germany, shows that humans hunted cave bears.

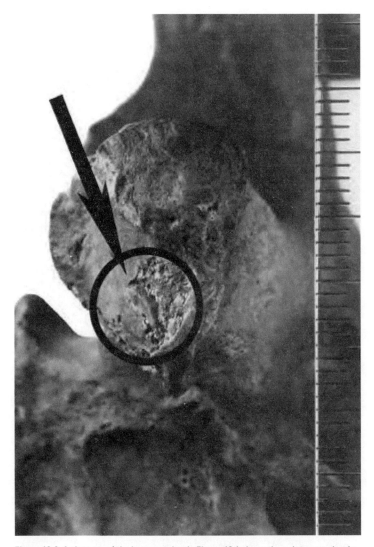

Figure 10.2. A close-up of the bear vertebra in Figure 10.1 shows the point more clearly.

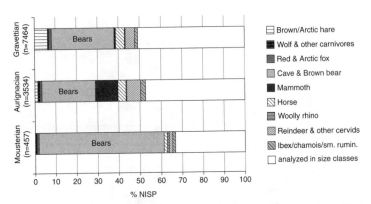

Figure 10.3. This diagram shows the proportion of bear remains (% number of individual specimens or NISP) in the fauna recovered from Hohle Fels from the older, Mousterian layer through to the Gravettian layer. Bear remains were more common in the Neanderthal (Mousterian) layer than in the modern human layers (Aurignacian and Gravettian). Most bears were killed in winter while they were denning, suggesting possible competition with hominins for favorable caves.

believes competition from modern humans was to blame, even though competition for meat was not at issue.[4] However, humans may have used bearskins along with other furs. And hominins needed plant foods, just as bears did.

At the same time that cave bear populations were plummeting, the density of archaeological finds (stone tools, refuse from tool making) in that region of Germany was rising, increasing by roughly a factor of ten to fifteen. This increase in archaeological materials after modern humans arrived is distinct. It echoes the pattern of rapid population increase of modern humans found by Mellars and French's study in France.[5] One would expect that the rise in human populations would increase the intensity of competition for any resources the bears shared with modern humans.

What was the crucial resource in this case? As Münzel and her colleagues summarize the situation, "The increased population densities of Aurignacian and Gravettian [modern human] hunters and the increased evidence for killing and butchering of cave bears strongly suggest that modern humans contributed to the extinction of cave bears via over hunting and by competition with bears for denning space and other resources."[6] The last known and well-dated western cave bear lived about 23,900 ± 110 (or between 28,730–28,500 calibrated years BP). Eastern bears survived a little longer, perhaps because they did not hibernate in such predictable and vulnerable locations. But they, too, went extinct after another 2,000 years. The bottom line is that although the strictly herbivorous cave bears did not compete with modern humans for prey, they did compete with modern humans in other ways—primarily for plants and very possibly for attractive cave sites.

The brown bear, *Ursus arctos,* survived after the cave bears were extinct and still survives today. Following the extinction of all competing cave bears, brown bears moved into the now-vacant dietary niche and became more herbivorous and less carnivorous. Of course, modern humans were still inhabiting the region and were formidable hunters and competitors. By using their dietary flexibility—a fundamental attribute of the bear adaptation—brown bears were able to escape the pressure of trophic competition with modern humans. Modern brown bears tend to den in shelters other than caves, which might have further spared them from intense competition with modern humans.

This well-worked-out synthesis of information on cave bear extinctions shows that the appearance of modern humans in the Eurasian ecosystems about 50,000 to 45,000 years ago had a far-reaching impact on the predatory guild. Competition was greatly heightened as the modern human population

swelled through time. Even cave bears, which did not compete with modern humans for food, were pushed into extinction through competition for denning spaces and through direct hunting by modern humans. During the bears' population
decline, the increased stress of climatic changes may have been
the coup de grâce for a species already in decline.

What about the other mammals that faced climate change
and human invasion at this time? When we look closely at
the Gravettian mammoth megasites and all the animals with
remains in those sites, we get hints of the peculiar pattern of
population declines or extinctions in various species at this
time. Though the bone assemblages from these sites are dominated by mammoths, many also feature large numbers of wolf
bones as well as many bones of Arctic foxes and hares; there
are fewer but still unusually high numbers of wolverines. All
of these animals have thick fur. Some of the bones of these
species with desirable pelts show marks from skinning. As
modern humans at this time had bone needles with eyes for
sewing, it seems self-apparent that some species were hunted
and skinned to make clothing as well as rugs to keep sleeping
humans warm.[7] Though tailored furs would have been an advantage in surviving in a colder habitat and in reducing metabolic needs, warm clothing might have made a significant
difference in the humans' ability to kill a mammoth if their
style of hunting mammoths required staying out in the cold
longer than, say, hunting wild horses, elk, or reindeer.

Nonetheless, killing animals for their fur—or making good
use of the fur of animals killed for food—was a new and important behavioral change for modern humans. Other changes
include scavenging mammoth bones from natural deaths on
the landscape for use in building and as fuel for fires. Still-

greasy bones of mammoths made excellent, long-burning fuel with a fairly steady temperature.

There is something else visible in the evidence. Some of the dead mammoths in modern human sites were killed and processed where they lay. "Of course," you may think, "nobody carries around a dead mammoth." Although some of the bones used to construct buildings at some sites seem to have been largely scavenged, others clearly were not. Some individuals are represented by whole carcasses. For one of the first times in human history, we see evidence that humans could not only kill a very, very large animal, but they could retain control of the carcass when challenged by other predators.

Many Gravettian sites include large numbers of carnivore bones. For example, Pavlov I in Czechoslovakia yielded 47,323 specimens, which came from a minimum of 536 individual mammals and birds.[8] The most common species was the hare, represented by nearly 200 individuals and nearly 6,800 specimens. But carnivore remains make up a remarkable 47 percent of the identifiable assemblage, which is highly unusual because carnivores are generally rare and fierce animals. In fact, the most abundant carnivore is the fox. (Arctic and red fox were treated as one species in this study because they can be difficult to tell apart from bony remains.) Foxes account for 123 individuals and more than 5,400 bones. There are also fifty-seven individual wolves (6,619 specimens) and ten individual wolverines (781 specimens). Reindeer, which are a more ordinary type of prey, are represented by fifty-six individuals and 4,000 bones, making them about as common as wolves. Because wolves are higher on the trophic pyramid than reindeer, we know they must have been much less common than reindeer in the living ecosystem. The abundance of wolves reflects targeted, preferential killing.

Grouse, ptarmigan, and ravens are abundant at the site and are represented by sixty-five individuals. Because bird bones are generally fragile, this is an unusual abundance. Their representation demonstrates the sort of increased dietary breadth that Kuhn and Stiner argue offered modern humans such an advantage over Neanderthals.[9] There are bones from a minimum of eight mammoths, which are represented not by all skeletal elements but mostly by those parts useful for making art objects or tools. Other species are poorly represented.

Of the specimens from Pavlov I, 330 show cut marks from skinning, dismembering, and removing meat. This is a remarkably high number because people who use stone tools do not want to blunt their tools by hitting bones. In comparison, only twenty-three bones from Pavlov I show marks from gnawing by carnivores, suggesting that carnivores had limited access to these remains. Somehow, they were kept at bay.

Another valuable resource derived from dead mammoths was bones and tusks, along with bones, antlers, and teeth from other mammals. At Pavlov I and other large mammoth sites, these raw materials were made into art objects, tools, and pendants. The abundance of carnivores is so unusual in archaeological collections that they might be disbelieved were they not echoed at other sites like Krems-Wachtberg in Austria, Dolní Věstonice I and II in the Czech Republic, Kostenki in Ukraine, and Kraków-Spadzista Street E and F in Poland.

From the onset of hominin hunting, defending a carcass from scavengers posed a major problem and was a serious danger. A basic hominin strategy was to use sharp stone tools to whack meat off of a carcass and carry it away to someplace safer, where potential competitors were less likely to attempt to take away the carcass. Neanderthals may have done just that to the mammoths at Spy, where cranial parts that would have been associated with fatty and nutritious mammoth brains were most numerous. At Molodova I, shoulder blades, crania, tusks, and pelvises—when largely clean of flesh, all useful bones for building mammoth bone huts—are favored by Neanderthals.

The big increase in mammoth-hunting success must be attributed to modern humans; a woolly mammoth represents a food bonanza. Gary Haynes of the University of Nevada, Reno, suggests that an adult mammoth might have yielded 500 kilograms of meat.[10] I think this might be an underestimate, because the meat weight from most living mammals is about 60 percent of total weight.[11] That percentage indicates that an adult mammoth would yield almost four times as much meat as Haynes estimated. However much usable meat and fat was yielded by a mammoth carcass, because the climate was cold, the mammoth meat might be able to be frozen in place for the winter months. Think of it as a natural freezer full of food.

Let's revisit the comparison between the return of a top predator (the wolf) to Yellowstone and the situation in Eurasia 50,000 years ago. When wolves returned to Yellowstone, they benefited other predators because wolves provided many more scavengeable carcasses for the ecosystem. Wolves' closest competitors, coyotes, suffered high predation by wolves, which was partly offset by the scavenging benefit.

This mixed benefit and disadvantage may not have occurred in the Paleolithic era, because modern humans found a way not only to kill mammoths and other large beasts but also to retain control of the carcasses. Neanderthals and other indigenous predators would not have reaped such a marked scavenging benefit from the presence of modern humans. The main impact of modern humans would have been heightened competition, which might either have decimated Neanderthals—as wolves did coyotes—or forced them into spatial separation and avoidance of humans.

One way in which invasive apex predators influence an ecosystem is by direct killing of the closest indigenous competitors (like wolves killing coyotes in Yellowstone). Logically, the closest predator to modern humans would have been Neanderthals, but there is relatively little direct evidence that modern humans were killing them. Although it has been suggested that modern humans may have killed Neanderthals and even eaten them, this interpretation is supported by very little solid evidence.[12] Modern human sites preserve remains of many prey species and in some cases of fur-bearing rival predators but very rarely any Neanderthal bones. Why not?

One possibility is that the two hominins did not compete as closely as the dietary studies discussed earlier suggest, though it is difficult to see where such a consistent error might arise. Another is that modern humans and Neanderthals had a limited overlap in time and space—no more than a few thousand years—and we simply may not have enough sites with good enough preservation to detect such evidence. Alternatively, Neanderthals may have soon learned to stay away from modern humans, or they may simply have died out rapidly given increased competition. Remember, Neanderthals hunted the same prey as did modern humans and as did nearly all of the other predators alive in this period. But Neander-

thals suffered from a limited ability to keep warm, a high metabolic rate, a vigorous lifestyle, handheld weapons, and the
loss of much of the wooded habitat in which they were most
successful at hunting.

In addition to killing and exploiting many more mammoths, and retaining control of their carcasses, modern humans at this time began targeting wolves, which rarely had
been killed by Neanderthals. The commonest explanation is
that wolves, like Arctic hares and Arctic foxes, were being
sought out for their dense fur, which certainly would have
been useful. But every dead mammoth would have yielded a
huge quantity of thick hide covered in hairs that are, on some
parts of the body, up to a meter long. Mammoth fur is coarse
and not as lush as wolf or fox fur, it is true. Are we looking
at the beginning of luxury goods? Was mammoth fur good
for making rope for nets or bags? Or is there some other factor
that made the killing of such fur-bearing animals—
particularly one of humans' fiercest competitors that was dangerous to challenge—suddenly become so possible and so important? Like the killing of mammoths, maybe the preferential
killing of wolves tells us something vital about modern humans at the time.

11

THE JAGGER PRINCIPLE

Some years ago, I heard ecologist Eös Szathmary refer to the Jagger principle: You can't always get what you want, but if you try sometimes, you just might find, you get what you need. I was immediately struck by the aptness of this remark. In some ways, the immortal words of Mick Jagger and Keith Richards are the best statement I know of to describe evolution.[1] Things don't stay the same; you can't always get what you want; but with a little flexibility, you might get what you need to survive. Brown bears managed this trick; cave bears and Neanderthals did not.

Some very strange things changed between about 45,000 and 35,000 years ago and, in total, are so extraordinary that they indicate a huge shift. To continue to get what they needed, most animals needed to change, too.

Let's review some of those changes. About 45,000 years ago, the climate began to fluctuate rapidly, with much colder and drier periods interspersed with warmer, wetter ones. The colder periods brought in the mammoth steppe fauna, with more huge, cold-adapted species. The dominant grazing herbivores were woolly mammoths, horse, and bison, but browsing or grazing herbivores like woolly rhinos, saiga antelope, reindeer, and other types of deer were present, too. Reindeer—judging from their modern counterparts and from the isotopic analysis of their bones—fed on grasses and supplemented their diets with leaves of willow and birch trees, but they also relied on lichens as a winter food, much more than did any other herbivore. The carnivores continued hunting: cave lion, cave hyena, cave bears, leopard, lesser scimitar cat, wolves, and smaller species like wolverines, dholes, Arctic foxes, wild cats, and weasels. Neanderthals, as adept hunters, lived on in the habitats that suited them best, the more wooded areas and the ecotones or transitions between woodland and open country. With changes in climate, the abundance and distribution of habitats also changed. Woodlands and forests contracted, then broke up into a patchier mosaic distribution, leaving large open swaths of steppe and tundra in between with wooded areas of birch, pine, and aspen along river courses. All this had happened before in previous iterations of the global cycle of glacial and interglacial periods. Populations had adapted, moved, or changed to survive.

But this time, there was a new apex predator, modern humans, who increased greatly in population size and territory. Modern humans had not been in Eurasia for the previous climate fluctuations, but Neanderthals had. Meta-analyses like that of Paul Mellars and Jennifer French [2] and the synthesis of information from the Ach Valley in Germany by Nick Conard and colleagues[3], discussed earlier, suggest that modern humans arrived, thrived, and rapidly surpassed Neanderthals in number. Neanderthal populations may already have been diminishing in population size and genetic diversity in response to the shrinking of woodland habitats suitable for their style of hunting. After about 40,000 years, there were few or no Neanderthals left.

Between about 40,000 and 35,000 years ago, a new type of site began to be made—one that attests to a striking behavioral or technological breakthrough. The change was the development of a new and more effective method for hunting megaherbivores, particularly mammoths. Modern humans were responsible for these new sites.

As mentioned earlier, there were large Gravettian sites with many mammoths—what I call megasites—which dominated the faunal assemblage. Mammoth megasites feature circular and occasionally rectangular buildings built largely of mammoth bones, which were presumably covered with mammoth hides (see Figure 6.1). There were multiple hearths, and postholes were created by the erection of poles to support a covering probably of hides. These mammoth bone huts were often part of large site complexes including butchering areas, storage pits, and tool-making centers. What is most remarkable is the staggering numbers of mammoth bones concentrated in small areas. Individual sites (or stratigraphic levels within sites) yield up to about 10,000 bones of as many as about 150 individual mammoths. Though bones of other species are also present, the assemblages are dominated by mammoth remains, which may comprise up to 90 percent of the bones (see Figure 11.1).

How unusual are the densities of mammoth individuals? We can compare mammoths only to living elephants. This is a reasonable analogy because we know that body mass is closely correlated with the size of the territory a species needs (and also its reproductive rate, age at first reproduction, age at death, and many other fascinating traits), and modern African elephants and woolly mammoths were similar in body size. In

Figure 11.1. This photograph of part of the excavation at Kraków-Spadzista Street, Poland, shows how densely packed the mammoth bones were.

protected areas, modern African elephants live in densities from 0.14 to sixteen individuals per square kilometer, as recorded in Hwange National Park in Zimbabwe. By 1960, elephant densities in Hwange had become so great that culling was undertaken to reduce the elephant density; culling was stopped in 1995. In Hwange, areas where elephant deaths were concentrated near water sources during droughts showed carcass densities of up to 1.5 individuals per square kilometer.[4]

In contrast, at the Gravettian site of Kraków-Spadzista Street (recently dated to 29,000 and 28,000 calendar years BP), the density of mammoth individuals is approximately one individual per square meter in one region of the site. This is an absurd density, orders of magnitude higher than recorded at the elephant death sites resulting from drought in Hwange. The mammoths could not have stood so close together when they were alive. This high-density area is interpreted as the kill site for these mammoths and as butchery areas where the carcasses were cut into manageable pieces, filleted, and in some cases cooked. As the principal analysts of Kraków-Spadzista Street, Jaroslaw Wilcyzyński, Piotr Wojtal, and Krzysztof Sobczyk of the Polish Academy of Sciences, rightly stress, there are about 23,300 individual specimens from this 170-square-meter excavation, which makes the site "absolutely extraordinary and unique."[5]

Let's put those the numbers into context by using a few assumptions based on what we know of modern elephants. Mammoths were very likely to have been herd animals, probably mostly living in maternal herds composed of a female and perhaps five of her offspring. If this assumption is correct, or even close to correct, then the eighty-six dead mammoths at Kraków-Spadzista Street would represent fourteen such maternal herds. Judging from modern elephant densities, that means the bones preserved at Kraków-Spadzista Street probably account for all of the mammoths in an area ranging from thirty-three square kilometers, if they lived at high density, to 215 square kilometers, if they lived at low density. Killing eighty-six mammoths at once would leave an enormous piece of landscape effectively devoid of mammoths—which is one reason we have to consider that the killing of all these mammoths did not occur at one time but over some years.

Kraków-Spadzista Street is indeed an extraordinary site, partly because of the density of stone tools, bones, and other artifacts, because of the care with which it has been excavated, and because of the rigorous analysis of the remains. Other, similar mammoth megasites in central Europe do not boast such very high densities, but they are still impressive, with densities of individual mammoths that range from 0.06 mammoths per square meter to 1.4 mammoths per square meter.

There are at least thirty other mammoth megasites containing hundreds of stone tools in the Gravettian or Aurignacian tradition made by modern humans, dating from almost 40,000 ago to about 15,000 years ago. (Unfortunately, nearly all of these dates were determined years ago and so are probably inaccurate; they are likely to be too recent due to contamination.) The number of individual mammoths represented at these sites ranges from about seven or eight to 166. Nearly all are concentrated in central Europe, in today's Poland, Ukraine, Russia, Germany, and Czech Republic (see Figure 11.2). Most include evidence of several mammoth bone huts or other buildings.

As discussed earlier, only two mammoth megasites are tentatively attributed to Neanderthals: Molodova I, level 4; and Spy. Molodova is a mass-death site composed of fourteen to fifteen individual mammoths that is about 44,000 years old (if we take at face value a radiocarbon date run decades ago). It has been argued that Molodova I-4 represents the oldest use of mammoths as food and their bones as raw materials for building by Neanderthals,[6] though this is not universally accepted.[7] It is not clear whether the mammoth bones at Molodova I were derived from hunted animals or were scavenged from natural deaths. Another Neanderthal site, Spy, in Belgium, also boasts a considerable number of mammoths probably killed by these hominins, not modern humans. Nearly all of the mammoths at Spy are extremely young, less than two years old at the time of death. Judging from data

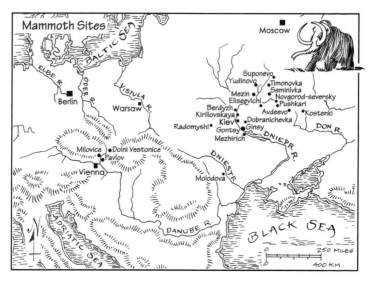

Figure 11.2. This map shows the locations of sites with mammoth bone huts, which may have been constructed of scavenged remains from natural deaths along the river valleys. Not all sites with many mammoth bones preserve signs of structures or dwellings.

on African elephants, these individuals were not yet weaned. Numerous samples from Spy have been recently dated using modern techniques, nearly all yielding dates between about 40,000 and 35,000 years BP and the redating project by the Oxford group suggests that last Neanderthal remains were close to 40,000 years old.[8] Possibly this was one of the last Neanderthal sites ever made. Isotopic analysis suggests that mammoths, in particular, contributed more to Neanderthal diet than an analysis of prey remains suggests.[9]

These mammoth megasites indicate that at least modern humans and perhaps Neanderthals, too, were capable of killing mammoths efficiently though they had not always been able to do so. Whether the new ability to kill mammoths is related to changes in the mammoths themselves is not clear, but a recent study of the mtDNA of 320 mammoth specimens from Belgium through Siberia and into the northwestern Beringia region of North America suggests some interesting wrinkles in the story.

A brief review of worldwide mammoth evolution will help place this work in time. Prior to 120,000 years ago, one lineage of woolly mammoths known as clade I evolved in North America and then spread into Eurasia. Clade is a term used to refer to a group of organisms with a common ancestor. According to the lead author, Eleftheria Palkopoulou from the Swedish Museum of Natural History, climate warming about 120,000 years ago "caused [woolly mammoth] populations to decline and become fragmented, in line with what we would expect for cold-adapted species such as the woolly mammoth."[10] The mammoth populations in the colder parts of the American continent were cut off from those in Eurasia by rising sea levels that drowned land bridges between the two continents. The once-continuous mammoth population was divided into two main popuations, one in Siberia and one in the Americas (see Figure 11.3).

Over millennia, the separation into two populations with subsequent genetic divergence led to the existence of two clades. When a cold spell about 65,000 years ago re-exposed the land bridges, the American mammoth population (clade I) came back into Eurasia and met up with the Siberian population (clade II) within the same species. The two coexisted for some time.

At about 44,000 years ago, clade I expanded their geographic range westward into Europe; at about 40,000 years ago, the remnants of the Siberian clade II went extinct. Clade

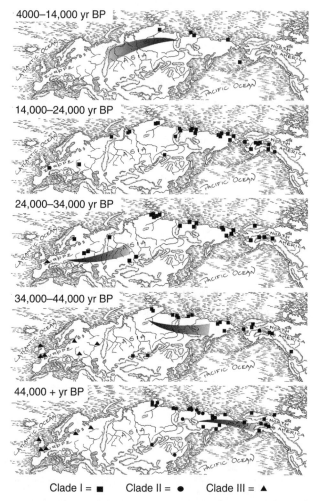

4000–14,000 yr BP

14,000–24,000 yr BP

24,000–34,000 yr BP

34,000–44,000 yr BP

44,000 + yr BP

Clade I = ■   Clade II = ●   Clade III = ▲

Figure 11.3. The migration, dispersal, or contraction of mammoth ranges over time has been reconstructed from genetic and fossil data. About 120,000 years ago, when temperatures were colder and drier, mammoths expanded from the New World into Eurasia. As the climate became colder, the distribution of mammoths moved west- and southward. When the climate became warmer and wetter, mammoths first retreated to the north of Eurasia. Eventually mammoths were confined to Wrangel Island, in the far north of North America.

I mammoths continued to expand their geographic range, reaching central and western Europe, where they met up with a group called clade III mammoths, which had probably been isolated in western Europe since the original mammoth expansion 120,000 years ago. Eventually, clade III mammoths also went extinct, about 34,000 years ago. The very last known representative of this clade was found at Goyet Cave, in Belgium, and is dated to about 36,000 calibrated years ago (or 32,280 ± 280 radiocarbon years).

As the climate continued to fluctuate rapidly during MIS 3, the remaining woolly mammoths of clade I survived only in areas that were farther and farther north. Eurasia was effectively devoid of mammoths by about 14,000 years ago except in the farthest northern areas, while Columbian mammoths lived on in the Americas. The last surviving mammoths anywhere were a dwarfed species that lived on Wrangel Island north of Siberia until about 3,000 years ago.

One of the authors of this remarkable study, Adrian Lister of the Natural History Museum in London, finds the work to be completely convincing evidence that climate change, rather than human activity, was the driving force behind mammoth extinction. I am not so sure. The disappearance of the European clade III lineage of mammoths happened shortly after the appearance of the mammoth megasites. The new and powerful style of mammoth hunting used by modern humans at these sites may have helped drive the clade III mammoths into extinction, given the huge numbers of mammoth carcasses involved in the Gravettian sites. And, of course, Neanderthals—which like clade III mammoths had been in western Europe since 120,000 or before—were already extinct.

Clades of mammoths and the Neanderthals certainly were not the only casualties of MIS 3. Many members of the carnivore guild crashed, too, though piecemeal and not simultaneously. This is not unexpected in a time of climate and environmental change. As a rule, medium- to large-bodied predators are unusually vulnerable to extinction because their survival depends on the survival of other species, their prey. Predators with a low reproductive rate—those that have on average less than one offspring per year—are even more vulnerable. The lesser scimitar cats and cave lions, cave hyenas, and dholes became locally extinct, and the cats and cave hyenas eventually were entirely extinct on the Eurasian continent. As a predatory hominin with high metabolic needs and a large number of competitors—and as a species that probably rarely had even one offspring per year—it is not so surprising that Neanderthals also went extinct. For comparison, adult female hunter-gatherers generally have one child every three to four years, though for many reasons hunter-gatherers—being modern humans—are a very poor model for Neanderthals.

The most notable exceptions to this crash in the predatory guild were the wolf and *Homo sapiens,* modern humans. Though being predatory puts a species at greater risk of extinction, one of these survivors had a very slow reproductive rate (humans) and the other generally had annual litters (wolves). Such litters were likely typical of many of the predators that went extinct locally or globally, like cave lions, cave hyenas, dholes, and probably lesser scimitar cats. Something crucial was at work here in addition to trophic level and reproductive rate.

Why did wolves and modern humans survive the tremendous faunal and environmental upheaval that decimated other species?

12

DOGGED

Once again, this part of the story of Neanderthal extinction and modern human survival begins with a remarkable and unexpected find. In 2009, a team led by Mietje Germonpré began publishing some startling discoveries. Interested in when and where dogs were first domesticated from wolves, Germonpré set out to look at the bony proportions of the skull that might distinguish dogs from wolves. First she and her team took standard measurements on skulls of forty-eight modern wolves, fifty-three modern dogs from eleven breeds (chow chow, Siberian husky, Malinois, German shepherd, Doberman pinscher, Irish wolfhound, rottweiler, Great Dane, mastiff, Tibetan mastiff, Central Asian shepherd), and five known and widely agreed upon prehistoric domestic dogs ranging in antiquity from about 22,000 to about 10,000 years BP according to older radiocarbon dates.[1] If the specimens used for these dates were contaminated, then they would actually have been older than these dates imply.

From the basic anatomical measurements, Germonpré's team constructed various ratios to express the proportions and overall shape of the skulls. Using complex statistical methods, they used this reference sample of known canids to identify sets of measurements that would distinguish one biological group from another. In effect, they constructed statistical pigeonholes or shape categories that represented each group.

The shape categories separated nicely with very little overlap. Even modern dogs with archaic skulls (chow chows and huskies), modern dogs with short tooth rows, and modern dogs with longer, more wolflike snouts could be separated statistically by the right measurements. Only one specimen, the Central Asian shepherd, lay completely outside the group that contained all other recent dogs. None of the modern wolves were misclassified as a dog of any sort. Importantly, the range of the prehistoric dogs fell completely outside the ranges of both the recent wolves and dogs. Because the separation among groups was so good, the team used the same techniques to measure eleven fossil canid skulls from various sites in Belgium, Russia, and Ukraine to see whether they fit into these pigeonholes. The statistical method they used, discriminant function analysis, not only classified each unknown into the most appropriate pigeonhole but also gave a percentage probability that this classification was correct, as well as a probability for the next most likely classification.

Their first publication on the subject spelled out their methodology and results. They announced that one of the eleven fossil skulls they analyzed—a large canid from the Belgian site of Goyet Cave—had a 99 percent chance of being a domestic dog rather than a wolf and was not matched by any canid in the modern reference sample. The Goyet dog fell between wolves and the sample of well-accepted prehistoric dogs (see Figure 12.1). This is the morphological space where an animal in the process of evolving from a wolf into a dog might be expected to fall. Two other fossil skulls they measured, one

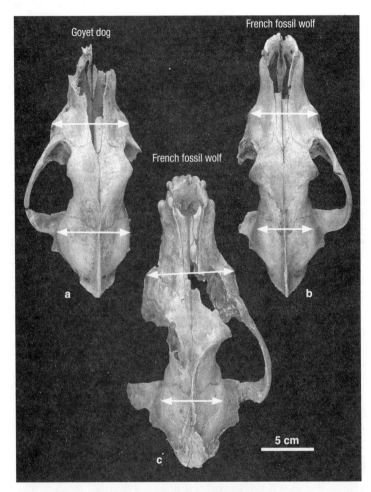

Figure 12.1. The first Paleolithic dog was identified from Goyet Cave in Belgium (*a*). Viewed from above, this skull is relatively wider in the braincase (lower arrow) than those of French fossil wolves (*b* and *c*). At 32,000 years old (uncalibrated), this specimen is much older than most scientists expected to find for domesticated dogs.

from a mammoth megasite at Mezin and another from Mezhirich, both in Ukraine, were also much more likely to be dogs than wolves, with 73 percent and 57 percent probabilities, respectively. These probabilities may not sound impressive, but the chances of those two Ukrainian dogs falling into the next-most-likely categories were much lower. These three skulls grouped close to each other but outside all of the predetermined pigeonholes. Seven of the remaining eight unknown skulls were classified as wolves, and one, from the site of Avdeevo, fit into none of the available categories.

These three specimens identified by Germonpré's team as dogs are also rather large animals, about the size of modern German shepherds. This is not particularly astonishing, as we know domesticated dogs were derived from wolves and Paleolithic wolves were large. But the sheer size of the animals might provide clues to their function—the "why" of their domestication.

The biggest surprise of all came when Germonpré had the newly recognized Belgian dog redated. Two samples, both carried out by the Oxford dating laboratory using modern techniques, gave dates of around 32,000 years ago uncalibrated (or roughly 36,000 calibrated years BP). This was unsurprising given that the archaeological evidence included Aurignacian tools made by modern humans. Prior to this study, the oldest well-accepted age of a prehistoric dog was judged to be at most 18,000 years old and more probably about 14,000 years old. Finding a dog 36,000 years old was unprecedented.

This is just the sort of discovery I personally like best. The methodology is sound, the comparative sample sizes respectable, and the result makes you rethink a lot of what you thought you knew. I have great admiration for this study, which was meticulously carried out and sensibly reported, without making too much of a surprising finding. In subsequent years, Germonpré and her colleagues have kept pursuing means of identifying fossil dogs and analyzing additional samples. I have been pondering what it means for our understanding of human evolution that the earliest modern humans domesticated dogs much sooner than anyone had anticipated, less than 10,000 years after humans arrived in Eurasia.

In a later study, Germonpré and her team identified three more fossil canids from the Czech site of Předmostí as being dogs, while two others (another from Předmostí and one from Kostenki) were identified as wolves, and three remained unassigned. [2] Still more recently, Germonpré and her team improved their anatomical and statistical technique for identifying early dogs further, using both crania and lower jaws or mandibles.[3] This is an important breakthrough because mandibles are more often preserved intact than skulls are, simply because they are less fragile. This approach has enabled the identification of still more early dogs from Předmostí, bringing the total to over thirty. I am sure the team will continue to expand their work to include specimens from additional sites.

"The Palaeolithic dogs in our data set are quite uniform in skull size and skull shape," Germonpré notes. [4] There is a stability in skull shape at the very beginning of dog domestication; these skulls look alike, and the animals they come from probably did, too. This finding helps confirm the distinctiveness of these specimens from wolves. As work has progressed and the sample of early dogs or wolf-dogs has enlarged, some interesting questions have arisen. A primary one concerns the identity of these unusual canids. Were they dogs? Were they wolves? For now, I call them "wolf-dogs" because it is not entirely clear which group they are affiliated with. But I do not mean to imply that these were hybrids of true domestic dogs and wolves, which are called wolf-dogs in the popular press today.

Olaf Thalmann, now of Turku University, Finland, trained as a postdoctoral fellow in the world-renowned canid genetics lab of Robert Wayne at the University of California, Los Angeles. Thalmann was able to extract mtDNA from an early fossil canid, called an "incipient dog," from Razboinichya Cave in the Altai Mountains.[5] Thalmann, Wayne, Germonpré, and a long list of geneticists and paleontologists coauthored a recent paper published in the journal *Science*.[6] The research offers additional genomic insights into the wolf-or-dog issue. The research team studied 148 canid mtDNA genomes, which included ancient mtDNA from eighteen prehistoric canids from Eurasia and the New World, which they analyzed along with complete mitochondrial genome sequences from forty-nine wolves, seventy-seven dogs—including "oddball" breeds such as the African basenji, the Australian dingo, and three indigenous Chinese dog breeds—and four coyotes. In thinking about the implications of their findings, we need to remember that mtDNA is carried only by the female line, not by males.

The resulting tree of genetic resemblances showed some interesting points (see Figure 12.2). Three fossil canids from Belgium, including the Goyet wolf-dog and two Goyet canids not identified as wolf-dogs, had the same highly unusual mtDNA. The mtDNA clade represented by these individuals has not been found elsewhere and does not appear among

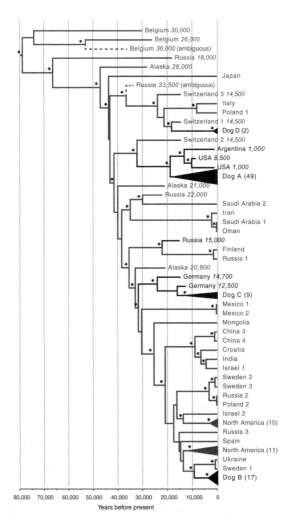

Figure 12.2. This tree shows genetic relationships among various fossil and living canids, based on their mitochondrial DNA (mtDNA). The mtDNA of the Goyet dog and another canid from Goyet (top three branches) are the same and are very primitive. Their mtDNA is not found in any other group. The four modern canid clades (Dogs A, B, C, and D) are shown by large dark triangles. Fossils are identified by country of origin and approximate age. Ambiguous classifications are indicated by "amb."

either fossil or modern dogs or wolves. This finding might mean that the wolf-dogs were descended from a group of local wolves whose females carried the same odd mtDNA haplotype. This lineage diverged from the ancestral forms long before any other canids yet sampled, on the basis both of the radiocarbon dates and of the distinctiveness of their mtDNA haplotype. These specimens span a time period from 36,000 to 26,000 years ago (calibrated). As the authors remark in the paper, these specimens are distinctive both in their genomes and in their appearance or phenotype. But what are they? "Given their mitochondrial distinctiveness, the Belgian canids, including the Goyet dog, may represent an aborted domestication episode or a phenotypically distinct and not previously recognized population of gray wolf."[7] The fossil canids identified by Germonpré's team as dogs (wolf-dogs here) on the basis of the statistical analysis of the shape of their skulls or jaws also have a distinctive mtDNA lineage or haplotype. But these specimens were not directly ancestral to any known living dog or wolf, because the mtDNA haplotype is not known from other canids.

Are these specimens dogs, then? Bob Wayne has been a friend of mine for thirty years, during which time he has become one of the preeminent researchers in canid genetics. We have been discussing these early wolf-dog specimens by e-mail for some years. He tends to call these specimens "wolves," not dogs. I asked him to explain why, and he replied, "Morphologically they were diagnosed as dogs, but if they are dogs, they should be directly ancestral to modern dogs. We now know they are not, because their mtDNA sequences are outside the radiation of dog and wolf sequences."[8] By this reasoning, of course, these canids are not wolves either, because their mtDNA does not fall within the known range of wolf sequences.

Deciding what to call these specimens without implying something we cannot justify is difficult, which is why I use the term "wolf-dog." They are a distinctive group, as several types of evidence have shown clearly. We do not know that they were wolves, and we do not know that they were the earliest dogs either.

The genetics suggest that sometime in the range of 36,000 to 26,000 years ago, humans may have succeeded in breeding a wolf-dog from a wolf stock but that these wolf-dogs were not directly ancestral to modern dogs, as far as we can judge with present samples. Though the Belgian mtDNA is not found in modern dogs, this does not rule out the possibility that this group was ancestral to modern dogs.

This is where the fact that the team is working with mtDNA becomes important. Male wolves with the unusual mtDNA could have contributed to the origins of the modern domesticated dog by mating with females from other mtDNA lineages. This is the same as saying that a male Neanderthal did not pass his mtDNA on to modern humans or to Neanderthal hybrid offspring, but nonetheless he could contribute to the modern human nuclear genome. Remember that Neanderthal mtDNA and modern human mtDNA do not overlap, but their nuclear DNA does.

One option is that the wolf-dogs might have been simply an odd group of wolves that arose and died out, leaving no female descendants. No specific geographic location where such a group might have evolved isolated from other wolves has yet been suggested. An advantage of this interpretation is that it does not involve postulating a surprisingly early domestication of dogs. However, this interpretation does not explain the peculiarities of the mammoth megasites, neither the new ability to kill large numbers of mammoths and retain control of the carcasses nor the sudden, targeted killing of wolves by humans.

How could a natural group of ancient wolves, with a separate mtDNA haplotype, evolve? This might be expected if a group of closely related females, sharing the same unusual mtDNA, moved into a new area with their offspring and a few outsider males. If this group then had little admixture with other wolves carrying different mtDNA, natural mutations might create and sustain an unusual haplotype for some time. Geographical isolation can occur because of landscape features or random events.

Another alternative involves arguing for the early domestication of a wolf-dog or doglike animal, on the basis of the archaeological and morphological changes in ancient canids. This interpretation would be the overwhelmingly most likely interpretation if the Belgian mtDNA haplotype were found among living dogs, but it is not. This raises the question of whether the genomic study was carried out on a large enough sample, worldwide, of both dogs and wolves to reveal all of the surviving (modern) mtDNA haplotypes. Also, because mtDNA is passed along only the maternal line (mother to daughter to granddaughter), most mtDNA lineages go extinct because no females with that haplotype had daughters.

How often mtDNA lineages go extinct—and how rapidly—is staggering. Evolutionary biologist John C. Avise of the University of California, Irvine, uses the example of the peopling of Pitcairn Island in the Pacific, drawing an analogy between surname survival (the male lineage) and the survival of mtDNA lineages, which are passed along the maternal line.[9] Pitcairn was settled in 1792 (following the mutiny on the HMS *Bounty*) by six male mutineers and thirteen Tahitian women. Assuming a human generation length was twenty years, only three of the original surnames survived after six or seven generations, plus one from a whaler who immigrated to the island later, in the population of fifty descendants. Similarly, Avise cites an example of a theoretical population with 100 different mtDNA lineages representing 100 women. After one generation, on average of thirty-seven of those original lineages have gone extinct because of random chance. After twenty generations, only ten of the original mtDNA lineages are expected to be surviving. Thus, the extinction of any particular mtDNA lineage in any mammal over thousands of years and generations is not particularly surprising.

The technical difficulty of the study by Thalmann, Wayne, and others should not be underestimated. The team was able to put together complete mitochondrial genome sequences on the reference samples and were able to extract ancient mtDNA from many more ancient canids than those discussed here. On the basis of their other genomic work, the authors are confident of the validity of their findings. After all, they typed seventy-two modern dogs and forty-nine wolf genomes from all around the world—more than anyone else had ever analyzed in canids. They feel that they have captured all or most of the genomic variability in wolves and dogs with these samples. Also, they used several different methods for clustering the genomes into phylogenetic trees, and all methods gave the same answers. The thoroughness of their work and sampling strategy implies that there are only four modern canid clades, referred to as Dog A, Dog B, Dog C, and Dog D.

These clusters leave the fifth ancient haplotype that turned up in the Belgian samples metaphorically out in the cold, closely related to neither wolves nor dogs. Of course, no one can be sure that this ancient haplotype does not lurk as a rare genome among living animals; a negative is hard to prove. This was one of the most comprehensive studies to date, but neither the dog nor the wolf reference samples in this study were as large as 100 individuals.

Part of the difficultly is that all living canids in the genus *Canis*—dogs, wolves, coyotes, Ethiopian or Simien wolves, and three species of jackals—can and do interbreed in the wild and produce fertile offspring. This means that even after dogs were first domesticated, there has been gene flow into the dog species from outside, complicating attempts to identify the exact origins of domestic dogs.[10] Another is that domestic dogs show unparalleled diversity in size, shape, and behavior, particularly in the last 200 years since selective breeding for dogs with different roles has occurred intensively. Dogs show an enormous range in size and skeletal proportions, greater than that in any other mammalian species. Diversity within domesticated dogs exceeds that of the entire canid family, and breeds may be retained as distinct genetic clusters because of regulated breeding.[11] In other words, one of the characteristics of the founder population—gray wolves—that was domesticated to create dogs was the potential for great morphological variability.

The Dog A clade encompasses a great deal of variability and 64 percent of the samples. It includes odd modern breeds such as Basenjis, dingoes, and two Chinese breeds, as well as three ancient pre-Columbian dogs that suggest that as people expanded into the Americas, they brought dogs that had been domesticated in the Old World. The Dog A clade is also closely related to an ancient wolf sequence from Kesserloch Cave in Switzerland dated to 14,500 years ago. The Goyet canids are most closely related to this clade but are nonetheless deeply divergent in mtDNA. The Dog B clade accounts for only 22 percent of the variability in domestic dogs and is most closely associated with modern European wolves from Sweden and Ukraine. The dogs in clade C comprise only 12 percent of the modern dog sample and are closely related to two morphologically distinct ancient dogs from Bonn-Oberkassel and the Kartstein cave in Germany (dated to 14,700 and 12,500 years ago, respectively). The Dog D clade shows the least diversity, including only two Scandinavian dog breeds. It is closely related to an ancient wolflike canid from Switzerland and then to another sequence from ancient European wolves, as well as extant wolves from Poland and Italy, and is rooted with the sequence from the putative incipient dog from Razboinichya in the Altai Mountains of Russia. Statistical analysis of the entire data set leads to three key conclusions, some of which have implications far beyond the questions I focus on here.

First, this work gives a strong indication that modern domestic dogs originated in Europe, because the earliest dogs to show a surviving mtDNA lineage are in Europe, not China or the Middle East, as has been suggested by previous, less comprehensive studies with smaller samples of mtDNA. This finding is important for understanding the location of dog domestication and for identifying the original domesticators.

Second, statistical analysis of the genetic data suggests that the origin of domestic dogs occurred sometime between 32,100 and 18,800 years ago. This age range spans the date of the ancient Belgian canids, including the Goyet Cave wolf-dog. Thus, the domestication of dogs may have occurred at the time of the Goyet wolf-dog and in the same general location.

Finally, the domestication of the dog occurred long before the domestication of other animal species or of agricultural crops, which began about 9,000 years ago.[12] This means that the humans involved in dog domestication were hunter-gatherers of some kind, not farmers. This fact refutes one of the most popular theorized scenarios of dog domestication. As elucidated by Ray and Lorna Coppinger, and repeated in many other publications, this scenario suggests that wolves self-domesticated into dogs by hanging around village garbage dumps and gradually becoming more tolerant of human presence.[13] But if domestication preceded agriculture, settled villages, and garbage dumps by thousands of years, then the ancestors of domestic dogs cannot have followed this pathway to cohabitation with humans.

Another problem with the Coppingers' scenario, according to animal behaviorist Valerius Geist of the University of Calgary, is that one of the key stages that eventually lead to wolf attacks on people occurs when wolves hang around garbage dumps.[14] Becoming familiar with humans, eating their refuse, and closely watching human behavior from a distance does not seem to lead to tamer behavior in wolves but to more aggressive behavior.

To me, the existence of this odd, distinctive group of wolf-dogs offers a key to understanding many of the unusual things that were happening in Eurasia in this period, including the extinction of Neanderthals. Nearly all of the wolf-dogs identified by skull or jaw morphology come from extraordinary mammoth megasites that attest to a breakthrough in hunting. To date, no archaeologist has interpreted the abundant remains of worked stone at these sites as containing a new tool or new type of technology that would foster more efficient mammoth hunting.

What exactly am I proposing? In its simplest form, I hypothesize that this odd group of wolf-dogs was indeed a first attempt at domestication, providing the advance that underpinned the formation of these mammoth megasites. The greater availability of food, particularly from fat-rich mammoth carcasses, in turn fueled the continued growth of modern human populations and territory. As modern human populations increased and became ever-more adept at taking prey, intraguild competition intensified. Neanderthals may have gone extinct in the early phases of the modern human invasion, but the expanded abilities of modern humans hunting with wolf-dogs—if they had wolf-dogs—would have triggered the extinction of most of the remaining predatory species. Assisted hunting may have contributed to the extinction of any last lingering Neanderthals as well.

One of the specific predictions generated by my hypothesis is that the first appearance of the mammoth megasites should coincide chronologically with the appearance of these genetically and morphologically distinctive wolf-dogs. Remember that, although modern humans in Africa had been hunting proboscideans since about 200,000 years ago, they had never encountered the combination of woolly mammoths and wolves until they entered Eurasia. The larger African carnivores theoretically available for domestication—such as jackals, Cape hunting dogs, Ethiopian wolves, hyenas, cheetahs, and lions—have never been domesticated. This was the first time in history that modern humans could possibly have domesticated a carnivore to assist them in hunting. There are no well-dated fossil sites with large numbers of dead mammoths older than the arrival of modern human invaders or older than the appearance of wolf-dogs, whether the appearance of dogs is demarcated by the statistical morphological analysis of dated skulls and jaws or estimated from the genomic data.

The problem is figuring out how all those mammoths died in the same places. While wrestling with this issue, I sought advice from Olga Soffer, professor emerita at the University of Illinois at Champagne-Urbana. She has studied the mammoth-bone-dominated faunas of Upper Paleolithic sites on the Russian Plain and is fluent in Russian. She warned me in an e-mail, "For God's sake stop buying the Hemingway myth that they killed every single mammoth—they settled near bone cemeteries and used them as lumber yards."[15] This was a wake-up call to me, and her humorous point is well taken. It is easy to be mesmerized by the large numbers of mammoths at these Gravettian sites and to forget that they do not necessarily represent kills from a single day, season, or year. Some of them could be natural deaths and many could be hunting deaths that occurred during reoccupation of a favorable site annually over decades. This issue might be resolved by obtaining modern radiocarbon dates on all individual mammoths used to make a single hut, but the only such dates are old and probably not reliable. It certainly requires great imagination to hypothesize how such large numbers of mammoths could be dispatched in a relatively small area in a relatively short period of time.

If leftover mammoth bones from old kills or natural deaths were seen as valuable building materials that would compensate for the relative scarcity of trees, Soffer's scenario is reasonably likely. However, recent excavation techniques have shown that the mammoth steppe was not utterly treeless, as was once suggested.[16] There were more trees—particularly birch and willow along waterways—than has been previously appreciated.

Mammoth carcasses provided advantages other than long, rigid bones and tusks suitable for holding up mammoth hide tents or mammoth bone huts. Mammoth bones and ivory were also used for making tools and art objects, and the direct killing of mammoths would not have been necessary if mammoth graveyards existed. Mammoth hides were huge, furry, and undoubtedly useful for covering mammoth bone huts, for making leather bags or ropes, as rugs to keep the floors of the huts warm, and so on, if the hide could be obtained before it rotted or was eaten by hungry predators. Using mammoth hides suggests that if modern humans did not kill the mammoths, at the very least they arrived at the scene of death shortly after the mammoth expired. Because the mammoth bones used in making huts are not generally chewed by carnivores, the humans also found a way to keep wild wolves, bears, cave hyenas, and other predators at bay. The use of mammoth resources does not prove mass killings, but sites with large numbers of mammoths need to be explained. Why would mammoths start dying in large numbers only after modern humans arrived? S. V. Leshchinsky and colleagues have suggested that woolly mammoths were stressed by nutritional deficiencies because of climate change,[17] but why such stresses might become acute after the arrival of modern humans is unclear. In any case, proboscideans are highly mobile animals and even in good times range over huge territories. We know humans killed at least some mammoths because more than a few mammoth megasites have yielded mammoth bones with cut marks suggestive of killing and butchery.

With the onset of the Upper Paleolithic industries— starting with the Gravettian in central Europe—the sites have an entirely different character from older ones. Though clearly all sites with lots of dead mammoths do not reflect human hunting and nothing else, it is an inescapable conclusion that human hunters played a vital role in creating the apparent boom in such sites.

In addition to the chronological coincidence, my hypothesis predicts that we ought to be able to see how having canid allies would help modern humans to kill the largest animal on the landscape, in numbers, and retain the carcasses. If wolf-dogs were sufficiently domesticated to work with humans, then sites where modern humans used wolf-dogs as hunting aids should show evidence of the clear advantages offered in modern times by hunting with dogs.[18] These include getting more meat faster and with less effort.

Modern dogs locate prey fairly quickly and hold game in place, by barking at and charging the prey while humans come up to kill it. For example, Finnish scientists Vera Ruusila and Mauri Pesonen undertook a study to see how much difference dogs made in hunting moose (the large cervid with palmate antlers that is called, in Europe, the Eurasian elk).[19] This is a traditional winter activity in Finland and is carried out on foot. Moose in Finland weigh 200 to 700 kilograms depending

Figure 12.3. A benefit of hunting with dogs is that they can locate prey quickly and hold an animal in place until the hunter arrives. Here, wolves from Yellowstone National Park use the same technique to hold a bison in place until it is tired and they can kill it.

on age and sex. They are very large animals, standing from 1.5 to over two meters tall. The dogs are large by dog standards—as are the wolf-dogs identified in the fossil record. Norwegian elkhounds or Finnish spitzes usually are used in elk hunting. They are sent off to find moose and keep them in place by surrounding them, barking, and charging until humans can approach and shoot the moose. Wolves can and do use similar tactics to capture large animals, though they do not work with human hunters (see Figure 12.3). In hunting groups of fewer than ten people, the average carcass weight per hunter without dogs was 8.4 kilograms per day. With dogs, the yield went up to 13.1 kilograms per hunter per day—an increase of 56 percent. Anything that improves hunting success by such a large margin is clearly a significant advantage.

Similar studies of modern hunting peoples in more tropical areas who use smaller dogs to hunt smaller game gave similar results. In Nicaragua, Jeremy Koster and Ken Tankersley of the University of Cincinnati studied hunting among the Mayangna and Miskito peoples of South America.[20] About 85 percent of all mammals were killed on hunts with dogs, which contributed to the catching of between 20 and 100 kilograms of meat per month—more than the dogs themselves weighed. The rate at which hunters found game improved with dog assistance by as much as nine times when the prey was an agouti (a large rodent). In a study among two tribes in the Central African Republic, Karen Lupo of Southern Methodist University found that dogs markedly reduced the time required to make a kill, by as much as 57 percent when porcupines were caught.[21] Though the specific hunting techniques, climate, and prey size varied enormously in these studies, hunters working with dogs clearly find more prey, find prey faster, and take home more meat.

These studies of modern hunters working with dogs suggest that ancient hunters working with wolf-dogs should have captured more prey, possibly a wider range of prey, and should have had an enhanced ability to take large prey while decreasing the human expenditure of energy in hunting. This is a reasonable description of what apparently happened at the mammoth megasites. Wolf-dogs provided a new way to exploit the ecosystem, by increasing the ability to take a wider range of prey and to take prey more reliably. Stiner and Kuhn remark that through more versatile methods of getting food, the carrying capacity of the ecosystem for early modern humans was effectively increased.[22] Though Stiner and Kuhn were not considering the effects of early domestication of dogs, hunting with wolf-dogs would surely have opened new opportunities and enhanced extant methods of obtaining food.

Another obvious advantage of working with wolf-dogs is reducing the human energy expended on the hunt while enhancing the success rate in terms of meat yield. As I said earlier, the wolf-dogs identified by Germonpré's team are large and strong animals. Writing of adaptations used by modern humans living in the Arctic, Charles Arnold of the University of Toronto explains, "Ethnographically-documented uses for domestic dogs *(Canis familiaris)* in northern societies include: drawing sleds, packing loads, locating breathing holes maintained by seals in the sea ice, holding muskoxen in their static defensive formation during the hunt, warning of camp intruders, and serving as a source of fur and food. Accordingly, domestic dogs played a significant role in the adaptive strategies of most historic Inuit and their archaeological predecessors, the Neoeskimo."[23] If they were trained not only as hunters but also as pack animals to carry loads, domestic dogs would be exceptionally valuable when the hunters did not want to camp at the kill site, perhaps because it was boggy, steep, or in some other way uncomfortable. Ethnographic data on the use of large dogs among Native Americans shows that dogs can drag travois or carry backpacks of up to about twenty-three kilograms.[24] The late Christy Turner, formerly of Arizona State University, argued that the invention of the needle and use of dogs enabled the eventual colonization of North America.[25] Another archaeologist, Stuart Fiedel of the Louis Berger Group, proposed that the earliest Native American women would have saved so much energy from the dogs' ability to carry firewood and food that their fertility might have been improved.[26] We do see evidence of an increase in the size of prey commonly taken and of an increasing population size among modern humans, as the benefits of having dogs suggest would have been likely.

The central nature of dogs to the adaptations of Inuit and far northern people, as expressed by Turner and others, has been challenged somewhat by Darcy Morey and Kim Aaris-Sørenson.[27] Though dogs were essential to the historical peoples of the far north that used dog teams to pull sleds and in hunting, this is a phenomenon of the last 1,000 years, they claim. Earlier peoples of the Arctic may have used individual dogs as pack animals or hunting assistants, but they did not have the big teams of six to eight dogs and the heavy sleds that later became ubiquitous. This explains the relative lack of dog (or wolf) remains in archaeological sites of the early Dorset culture when compared with those of the Thule, where not only dog remains but also pieces of harness or sled are commonly found. Sites of the Dorset people (Palaeoeskimos) precede the Thule culture and do not yield numerous dog or canid remains, though dog remains are present in small numbers, and some bear pathologies that may be related to hauling.

Another valuable role such wolf-dogs might have played is guarding the carcass and defending it from other predators. Dogs are still used today in the Arctic to warn people of the approach of wolves or polar bears. Dogs are used by people armed with bows and arrows to hunt polar bears as well as caribou (reindeer) and seals.[28]

One of the distinctive features of the mammoth megasites is that in many cases there is strong evidence that modern

human hunters camped on or very close to the kill site because they were better able to retain control of carcasses and
did not have to flee to safety. The combination of reduced
hunting time, increased hunting success, lessened energy expenditure, and enhanced carcass retention may have spelled
the difference between an occasional mammoth kill and the
killing of mammoths on a large scale as first practiced by early
modern humans about 40,000 to 32,000 years ago.

Stiner and Kuhn have suggested that hunting in the Paleolithic was a predominantly male occupation.[29] If they are
correct, then wolf-dogs might have played a crucial role in
defending women and children from marauders while the
men were out hunting. These marauders could have been
simply outsiders or wild animals. Such protection might have
been vital to the reproductive success of modern human
groups.

There might have been some other interesting consequences
of working with these wolf-dogs. Recent behaviorists studying
wolves have found that wolves target other canids—
domesticated or not—preferentially. Constant challenges to
or attacks on wolf-dogs allied with humans might well have
motivated humans to kill wild wolves whenever possible.
Wolves are also fiercely territorial. If modern humans lived
and traveled with wolf-dogs, local wolves would have been
extremely aggressive toward these newcomers and competitors. Ancient wolf-dogs were probably vigilant about the possibility of a wild wolf pack approaching. One way of thinking
about this is that if modern humans were allied with wolf-
dogs, they would have needed to offer some protection to their
canids when they traveled. Domestication is always a two-
way street, a negotiated arrangement that must benefit both

partners. Working with wolf-dogs would have made wild wolves a more dire and formidable competitor to modern humans than they had been previously.

Abi Vanak of the Ashoka Trust for Research in Ecology and the Environment, Bangalore, and Matthew Gompper of the University of Missouri looked at the interactions between dogs and wild carnivore populations. They found that the dog–human alliance has a beneficial influence on the interactions of dogs and wild carnivores, from the dog's point of view. Dogs are less subject to food scarcity because of human handouts and are protected from attack or competition from other carnivores. Human habitations provide safe refuges.[30]

Let's consider the archaeological data from a few sites in light of my hypothesis and its predictions. Mentioned earlier, Pavlov I is a Gravettian site in the Czech Republic. Pavlov I yielded bones of seven individual mammoths, as well as the remains of fifty-seven wolves, ten wolverines, four bears, and 123 arctic foxes. Each bone was carefully scrutinized for cut marks, and a judgment was made whether the carcass was skinned, filleted or defleshed, or disarticulated according to its location and modern hunters' practices. Three hundred and thirty-two bones in the assemblage showed cut marks made during the skinning of these animals; these marks were commonest on reindeer. Piotr Wojtal of the Polish Academy of Science in Kraków and colleagues write: "The large number of carnivore remains and numerous signs of skinning show that they provided hides for clothing preparation. Skinning cut marks were found on 22 carnivore bones, mostly [on the tail vertebrae and on the bones of the paws]. They show that hunters tried to get hides of the animals as large as possible. We have to point out the presence of the cut marks made

Figure 12.4. Many clay fragments from Dolní Věstonice and Pavlov show the impression of fibers made from vegetable matter or fur. Some show sophisticated knotting or weaving, suggesting that netting, baskets, or other items were made and used.

during dismembering and filleting of wolves, foxes, wolverines, and bears, indicating that carnivores were also a source of food for the Pavlov I hunters."[31]

The nearby Gravettian site of Dolní Věstonice is famous for its wonderful ceramic Venus figurine, but much more was excavated there. Mammoths, wolves, and foxes are three of the most common species in the fauna. Both Pavlov I and Dolní Věstonice also preserve small pieces of clay marked with the impressions of cordage and woven materials—the oldest in the world at 27,000 to 25,000 BP (see Figure 12.4). The cordage was almost certainly used in snares or nets to catch medium or small animals, possibly in butchering or carrying prey, and in constructing the mammoth-bone houses. This fiber technology was also a part of the new set of adaptations that gave modern humans the edge over Neanderthals—but fiber alone cannot have been the breakthrough that enabled modern humans to create mammoth megasites.

At Předmostí, faunal analysis reveals the dominance of mammoth bones, accounting for more than 1,000 individuals. The next most important animal is wolf, with 103 individuals, followed by arctic hare (seventy-nine individuals), and reindeer (thirty-six individuals).[32] Twenty canid individuals have been identified as wolf-dogs, although fifty-two other specimens were too incomplete to be identified by the techniques used by Germonpré's group.

Following up on these findings, Hervé Bocherens and colleagues carried out an extensive isotopic study of the entire fauna from Předmostí, sampling virtually all the medium- and large-sized species represented there: reindeer, bison, red deer, muskox, horse, woolly rhino, woolly mammoth, brown bear, wolverine, arctic fox, cave lion, wolf, wolf-dogs, and three modern human individuals who were buried at the site.[33] After eliminating bones too poorly preserved for analysis—as judged by the content of collagen, the major protein in bone—the group obtained remarkable results. The stunning conclusion is that the cave lions ate great quantities of reindeer and bison, while the wolves ate more horse and perhaps more mammoth. But the large canids interpreted as wolf-dogs did not have isotopes that matched those of most wolves; instead, they seemed to be relying more heavily on reindeer for food. The humans from Předmostí and the Moravian sites nearby show isotopes consistent with eating large quantities of mammoth, so if they were provisioning the wolf-dogs, they were giving them not leftovers from their own meals but reindeer. Could they have been hunting reindeer specifically for feeding the wolf-dogs? Or, as the freshness of the reindeer deteriorated over time, was it downgraded from human food to wolf-dog food?

The presence and assistance of wolf-dogs would explain the enhanced ability of modern humans to obtain and retain large carcasses and also their intentional targeting of wolves as prey. Together with the predictions based on invasions by predatory species, this scenario makes sense of the sudden changes in modern human hunting success and their ability to survive in an environment that, by all accounts, was formidably difficult. Modern humans may have survived simply because they were dogged.

13

WHY DOGS?

If the unusual wolf-dogs identified at various sites represent an early attempt at domestication—a previously unknown and far-from-obvious endeavor—can we explain why early modern humans chose dogs as the first animal to be domesticated? Many thinkers have pointed to common traits among animal species that have been domesticated.[1] Do wolves fit the bill?

First, animals that are social are good candidates for domestication. Wolves are the unquestioned starting point for the domestication of the dog and are very social animals that live in packs with a hierarchical social order. Wolves already have skills and adaptations for living in groups, hunting cooperatively, and establishing which animals are dominant over others. They also share care of offspring with other group members. Thus, if humans can be accepted as pack leaders, the structure of the interspecific interactions will be familiar.

Second, rapid reproduction and rapid growth to maturity are desirable traits, as this means more of the domesticated animals will be produced in a shorter time. Willingness to breed in captivity is often cited as another characteristic, though what "captivity" might mean in the case of wolves being domesticated 35,000 years ago is open to real question.

In any case, compared with other, later domesticates like goats, canids are a poor choice in this regard.

Third, an ability to thrive on food not eaten by or not prized by humans is helpful. Thus, a stock animal that can feed on grass or crop stubble is an obvious candidate for domestication because feeding them does not take food out of humans' mouths. Wolves or dogs, which compete directly with humans for meat, are not obvious candidates for domestication in this regard. The benefit of domestication or alliance or cooperation has to be greater than the cost of giving food to the domesticate.

Fourth, likely candidates for domestication will have a calm temperament, will not be very aggressive, and therefore will be fairly tractable. Wolves do not fit these criteria.

A fifth trait, not generally cited as a characteristic of a good candidate species for domestication, is that wolves and humans share an occupation, a means of getting food. Wolves do not have to be taught how to hunt in groups by humans; they learn from their parents and, to a real extent, hunt instinctively. This trait applies to some extent to domesticated cats, which knew how to hunt rodents and other small mammals long before humans domesticated them sometime between 3,800 and 1,600 years BC. Whereas wolves and dogs hunt cooperatively, felines tend to hunt in a solitary fashion and do so as domesticated cats. A hunting dog works with humans, a hunting wolf hunts with other wolves, but a hunting cat hunts alone.

The reason the first species to be domesticated, the wolf, fits these criteria so badly is that the criteria are based on the underlying assumption that domestication is "about" ensuring a meat supply by raising stock animals. The model animal for domestication has traditionally been something more like a sheep or goat than a wolf. Two highly influential papers on animal domestication were written in 1981 and 1983 by Andrew Sherratt.[2] In these works, Sherratt put forward the hypothesis that animals were domesticated primarily to provide food security and only secondarily and later did humans systematically exploit the other secondary products animals provide, such as milk, wool, or traction. The impact Sherratt's work had on anthropologists and other scientists who study domestication is huge.

Unfortunately, I think Sherratt was dead wrong. I argued in a recent paper[3] that his hypothesis is based on a fundamental misunderstanding of the motive for domesticating animals. If the aim was to produce meat, then domestication is a poor strategy because the process inevitably takes a considerable number of generations before someone can walk into a paddock or fenced area and select an animal to kill for dinner. I realize that some domesticated animals have proven to be useful food sources; that point is obvious. But long before they were seen as food, various species offered many renewable resources (a better term than "secondary products") that do not require killing the animal. These renewable resources include milk, fur, power for traction or hauling, assistance in hunting, protection, sanitation (eating refuse and other substances inedible to humans), and the ability to make more animals. A pertinent example is the remains of knotted, spun, and dyed flax fibers and mountain goat hairs that date to between 32,000 and 26,000 years ago recovered from Upper Paleolithic layers in Dzudzuana Cave in the Republic of Georgia. There are also numerous impressions in clay of spun and knitted or woven fibers from Pavlov I. These finds are direct evidence that the use of animal fur and plant fibers had reached a significant level of sophistication well before the domestication of the most obvious wool-bearing animals, sheep and goats, at about 9,000 years ago.

I suggest that the original reason for domesticating animals, as exemplified by the wolf-to-dog transition, was to create "living tools," so that humans could borrow the useful attributes of the domesticates that were not shared by humans. From wolves and dogs come the ability to run swiftly over long distances, to track by scent, or to hold prey at bay by surrounding, growling, and, if necessary, charging or biting the prey. If a species with these traits and abilities will work with humans, both can benefit by obtaining more meat faster and more meat per hunter (human or canid), with smaller risk of injury, among other advantages.

It is worth stopping here to consider what exactly domestication entails. Domesticating a species is not at all the same as taming an individual animal, which in itself can be difficult. Taming involves setting up some behavioral rules—like "don't chew on the children" or "don't steal my food"—and enforcing them until the wild animal knows and accepts the rules or is thrown out or killed. This sort of training is usually easiest to accomplish by starting with a young animal. With wolves, there is a period of socialization during which pups explore the outside world without fear, learning what is safe. It occurs starting at about two weeks of age and continues for a month.[4] This is obviously the prime opportunity to begin taming an animal. But taming does not change the species genetically, and each generation is as wild as the last. The behavioral changes are not passed on to offspring.

Both taming and domesticating a species may require a considerable degree of empathy and understanding and a well-developed ability to comprehend the significance of various postures, gestures, and noises that the potential domesticate, or potential domesticator, uses to express friendliness, fear, hostility, illness, hunger, or curiosity. There is a long learning process in communicating with an animal, even if that communication is very rudimentary and trivial. Both animal and human must participate and seek to communicate with the other; communication must seem both possible and worthwhile. Television shows on dog training and dog whisperers demonstrate clearly how badly behaved even a domestic dog may be when raised by humans who do not know how to encourage appropriate behavior and discourage dangerous behavior. Many of these "failed" owners do not understand when the animal is angry, being dominant or aggressive, or is frightened. They don't "speak dog," and learning how to understand dog language is usually a key part of the program.

One of the ultimately important aspects of domesticating an animal is that people control the breeding of the potential domesticate and take care that individual animals with undesirable traits—such excessive aggression toward humans—do not survive. Domesticating may or may not be a conscious process. Indeed, no human seeking a family-friendly, useful dog, like most we know today, would start with a wolf. But the beginning of the process of domestication may be as instinctive as killing an animal that is too much trouble or too dangerous or does not behave as you want it to. In essence, domestication is about genetic changes that alter the species forever.

The most famous experiment in canid domestication was carried out in Siberia starting in 1959, at the Institute of

Cytology and Genetics in Novosibirsk, Russia, under the direction of Dmitri Belyaev. Silver foxes—an especially attractive color variant of red foxes with long, silver-white guard hairs—have been actively raised in Siberia for the fur trade since the nineteenth century because the cold climate makes their fur particularly lush and thick. Belyaev chose thirty male foxes and 100 vixens from an Estonian fur farm as his initial or founder population. Although these lineages had been bred in captivity for fifty years, Belyaev regarded them as "virtually wild"—which they were not, as they had been exposed to humans frequently. The most fearful or aggressive animals had already been weeded out over those fifty years.

These foxes lived in cages with minimal contact with humans except for a once-a-month behavioral test conducted on newborns to see how fearful or friendly they were toward humans. Breeding was strictly dictated by Belyaev's aim, which was to select the most friendly and least fearful foxes to cross with each other; another group was allowed to mate at random, as a control group.

Brian Hare, an expert in dog behavior at Duke University, visited Novosibirsk in 2003, after forty-five generations of selected silver foxes had been bred. He describes the control group as "melting away into the back of their rooms" to avoid him and vocalizing with a threat bark as he passed. The foxes bred for friendliness "leapt into my arms, nuzzled against my face, and licked my cheeks with their little pink tongues."[5] It is an engaging description. Only 18 percent of the animals were judged "fearful" in contrast to the 90 percent in the founder group. The resultant animals, which are often called "domesticated foxes," were far less fearful of humans than the controls, leading to the exaggerated claim that it takes only forty-five generations to domesticate an animal.

I think this conclusion is overly simplistic because the control Belyaev had over these animals is far more extensive than anyone could possibly have in any attempt at domestication under natural conditions, starting from the selection of less wild silver foxes through to the rigid choices of breeding partners. The foxes were kept in cages and had no choice of breeding partners. I regard forty to forty-five generations as representing the bare minimum amount of time needed to accomplish domestication, if the animal in question breeds annually and if the people in question have a modern knowledge of genetics and inheritance and extreme control over the animals.

It is fascinating that other genetic changes came along accidentally while Belyaev bred for friendliness. The silver foxes developed varied coat coloration, some showing up with piebald patterns never observed in the founder population; some had floppy ears instead of erect ones; some of their tails curled upward; their skulls were, on average, narrower, with shorter, wider muzzles. Most crucially, their basal levels of the stress hormones called corticosteroids were much lower than in the wild foxes. The youngsters also had a prolonged period of curiosity and openness to new things before fearful behavior set in.[6] This timing meant that training and achieving communication was easier in these specially bred foxes, probably improving with each new generation. There is no way to estimate either how long it might have taken early modern humans to domesticate wolves into wolf-dogs or dogs or how many times keeping a wolf puppy around failed and ended in the death of the puppy. My guess is that there were many attempts and many failures.

Of all the animals that might have been selected for domestication, a large, particularly ferocious one like a wolf seems a stupid choice.[7] There were other species in the Pleistocene fauna, such as goats or deer, which seem more docile and tractable. (Deer, however, have never been domesticated.) But domestication was almost certainly not in the minds of early modern humans when they embarked on the journey to domestication. I cannot envision the domestication of dogs happening in any but an accidental way, through trial and error that probably started with a wolf pup or two and which only occasionally turned out well. Because there were no other domesticated animals, it seems unlikely that anyone envisioned the benefits of a domesticated animal and set out to breed one. With the exception of dogs, all of the successful domestication attempts began much later in time, after humans were regularly living in settled places, often raising grains and gathering plant foods.

Still, wolves offered some advantageous talents and abilities as potential domesticates, despite their dangerousness. Wolves, like humans, are intrinsically social animals that live cooperatively in packs. They are happiest living in a group of their kin or selected associates rather than alone. A solo wolf is a wolf in danger, a wolf with a compromised ability to hunt, a wolf with no help to raise its young, and a wolf with no one to watch his or her back. In this essential regard, humans and wolves (or most canids) are very much alike. Enlarging the concept of family or pack to include a wolf-dog, or a human, is not beyond belief if the alien species was raised with humans and learned to read human signals. And wolves, dogs, and humans are all happier living in groups.

Wolves are also inherently hunters, and they are better at smelling, tracking, and chasing down prey than humans will ever be. Though their main means of obtaining food is by hunting, as among early modern humans, the techniques used by canids are in many ways superior. They can trot in pursuit of prey for twenty-five or thirty miles in a day without undue fatigue. They can follow a scent trail easily. They instinctively run down prey by alternating the lead position among members of the pack. What canids do not have and humans do possess are long-range weapons like thrown spears, atlatls or spear throwers to propel spears with great force, and bows and arrows. Humans can also construct snares or traps no canid could create to catch an animal with relatively little effort. Large canids do well physically attacking prey, but that approach can be dangerous to them and unsuccessful, too. Weakening an animal through loss of blood, immobilizing it in a snare, or wounding it and then tracking it until it can fight no more are much lower-risk strategies than a direct attack.

Wolves are said to be smarter than dogs and to work together cooperatively better than dogs.[8] Wolves also perform better on spatial tasks.[9] Valerius Geist refers to wolves as insight learners, who can observe other animals or even other species and learn how they solve a problem, such as unlatching a gate.[10] However, the relative intelligence of dogs and wolves depends on what the situation is. In studies, dogs pick up cues from humans that wolves ignore, and dogs look to their owners or handlers for permission to carry out specific tasks or for help in difficult tasks. As Brian Hare and Vanessa Woods write in *The Genius of Dogs*, "Dogs are one of the most important species we can study. Not because they have become soft and complacent compared to their wild cousins but because they were smart enough to come in from the cold and become part of the family."[11]

Thus, it is particularly intriguing that the bones of wild wolves—the carnivore that would have been the most competitive with modern humans after Neanderthals went extinct—are extremely common at sites where the early wolf-dogs and many mammoths have been found. Arctic fox and hare are also very common at these sites.

Why were the wolves and foxes captured at all? Cut marks and the presence of many cranial remains, along with bone of the paws, indicate regular skinning of these animals, as paws and cranial parts are often left attached to the skins. Thus, the general assumption has been that wolves and foxes, along with hares, were targeted for their furs. Fur unquestionably would have been very useful in cold periods. In older publications, many authors are adamant that wolf carcasses were not eaten. Olga Soffer expresses this conviction clearly in her 1985 book: "While numerous remains of fur bearers are found at most of the sites, the bones belonging to these taxa never show any signs of food processing or food preparation."[12]

But why not? Wolf meat is surely edible, people happily eat canids today in some parts of the world, and a wolf weighs about as much as a deer, for example. Why kill them and skin them and not eat them? I can find no sensible explanation other than some presumed cultural preference or significance. In the study of interference competition among carnivores by Palomares and Caro, about half of the species that engaged in intraguild competition ate their foes after killing them and half did not.[13] They could detect no characteristics that predicted whether the carnivorous competitor was or was not eaten. As Vanak and Gompper remarked, "when the subordinate species is not consumed after intraguild predation . . . this suggests that the dominant species is directly reducing numbers of the putative resource competitor."[14]

However, in some more recent excavations of Gravettian mammoth megasites, like Pavlov I, wolves, red foxes, and arctic foxes clearly show signs of skinning, dismembering, and filleting (meat removal).[15] Were these marks missed in older studies, when the importance of cut marks and butchery marks had not yet become so clear to zooarchaeologists as they are today? Or were dead carnivores simply not eaten in some areas of the Pleistocene world whereas they were eaten in others? We do not know.

The distribution of cut marks on wolves at Pavlov I matches fairly closely those left on bones of herbivorous prey species like reindeer, providing compelling reason to assume that at least some these canids were eaten. Possibly wolves were preferred prey because they were hanging around human dwellings, hoping to scavenge meat; killing some might discourage very unwelcome visits from others. Their remains were useful in other ways, too. Awls and other bone tools were regularly made from wolf ulnas from the lower foreleg; these bones are naturally elongated and pointed. Their teeth were also preferentially used to make pendants, by drilling a hole in the root of the tooth. At Pavlov I, for example, 254 fox teeth and sixty-five wolf teeth—a large number—were treated in this way. Only two such pendants from Pavlov I were made from another species, reindeer, though reindeer remains were much more common.

Unexpectedly, the bone assemblages at the Gravettian sites do not include large percentages of tooth marks from carnivore chewing. Many of these sites show carnivore tooth marks on fewer than 5 percent of the specimens. As carnivore-chewed bones are common at more recent sites with dogs or at wolf dens, why are they not common in these assemblages? There are three possible explanations.

First, procedures at many sites excavated earlier than the late twentieth or early twenty-first centuries often did not include collecting or analyzing all fossil remains, only the more complete ones. It was common practice to ignore scraps or fragments of bone, heavily chewed or not. Such collecting procedures make it impossible to know now whether fragments of uncollected bones exhibited gnawing or chewing marks.

Second, humans may have been successful at keeping wild carnivores away from their camps or settlements, even temporary ones. Retaining such efficient protection of carcasses would have been a formidable task without the help of wolf-dogs. But if wolf-dogs were cooperating and living with humans, keeping them from chewing bones would have been exceptionally difficult—even if they were well fed—unless they were tethered or otherwise confined.

Finally, the extent to which humans or carnivores process bones is dependent on the relative amount of meat available to them and the duration of the occupation of the area. Therefore, carcass parts brought back to the den or to animals confined in enclosed areas are much more likely to be heavily damaged by chewing than those at the kill site. Indeed, sites where wild wolves have killed a prey animal may consist of no more than a handful of bone fragments, none bearing tooth marks.[16]

Mary Stiner has studied bone damage patterns that distinguish ancient and modern wolf dens from cave or spotted hyena dens and from hominin-collected assemblages.[17] What she defines as a carnivore den assemblage is one marked by the presence of bones from cubs, one with heavy gnawing damage to the bones in the assemblage, and, in the case of cave hyenas, many coprolites (fossilized droppings). In other words, carnivore den assemblages occur in maternal dens where carnivores probably bore and raised their young for periods up to several years among some species. The "heavy gnawing" Stiner specifies takes the form of tooth marks on roughly 10 to 50 percent of the bones. Another distinctive feature of carnivore dens is a high proportion (up to 70 percent of the number of specimens) of carnivore bones. Probably because of inter- and intraspecific competition, carnivores tend to kill many other carnivores. Hyenas tend to produce assemblages dominated by head and antler specimens, while wolves more often collect limb bones. Assemblages produced by hyenas, wolves, and hominins also produce different patterns in terms of the age structure of the prey animals. Hyenas and wolves favor juvenile animals, while hominins favor prime adults.

In Stiner's sample, carnivore den assemblages have few or no stone tools, no burned bones, and no cut marks or percussion marks indicating damage made by hominins. Thus, these den sites with lots of carnivore tooth marks and damage differ markedly from the mammoth megasites in Eurasia, which show damage from burning, skinning, filleting, and disarticulation of animal carcasses plus lots of stone tools or debitage (small pieces of stone that are a by-product of knapping). The mammoth sites usually lack juvenile carnivore bones. In summary, a low degree of carnivore-chewed bones is to be expected at these Gravettian sites, where human presence was abundant, where meat was plentiful, where wolf-dogs or dogs were not breeding, and where wolf-dogs, if present, may not have been confined.

Do we have solid evidence that wolf-dogs were a special, perhaps partly domesticated species? If preferentially killing canids and making jewelry out of their teeth are signs of special treatment—or special status—then we have strong evidence. But there are hints and suggestions of other kinds that support a certain amount of speculation about their status in the eyes of early modern humans.

This speculation would seem to be ill founded except for the evidence of much later finds. In a consideration of the worldwide evidence, Darcy Morey, a specialist in dog burials, remarks, "Nothing signifies the social importance that people have attached to dogs more conspicuously than their deliberate interment upon death."[18]

There is evidence from more recent sites—about 14,000 years ago to the present—that people from cultures worldwide have deliberately buried dogs and even sometimes included grave goods in the burials. No other animal species has ever been so often treated in a ritual manner as dogs have been. One example is Ashkelon, a site in Israel dated to between 2,500 and 2,200 BP, where 1,000 dogs were individually buried in a canine cemetery area.[19] In the Cis-Baikal region of East Siberia, domesticated dogs were buried with humans in the same cemeteries and sometimes in the same graves during the early Neolithic period (8,000–7,000 BC) and in the early Bronze Age (5,000–3,400 BC). The most remarkable example is from the human habitation site of Ust'-Belaia, where a dog was buried wearing a necklace made of eight canine teeth of red deer and with a bovid horn core and shoulder blade, two antlers of roe deer, and other identified bones. Burial of dogs with humans in this region seemed to occur only among foraging peoples with a substantial reliance on aquatic foods.

In 1894, Karel Maška, one of the early excavators of Předmostí, reported in his diary that he had found seven or eight complete wolf skeletons with partly broken skulls.[20] One of these was found in the area where twenty human skeletons were also interred. The evidence is highly suggestive that the covenant or alliance between the wolf-dogs and modern humans made the humans treat human and wolf-dog remains in similar ways. Possibly wolf-dogs were viewed as being too close to humans to be treated as other animals were.

But does the inclusion of a canid in an area of Předmostí where humans were buried suggest this animal was considered to be near humans about 30,000 years ago? Unfortunately, Maška did not record the spatial position of each piece, so we do not know which find came from the human area. If that deliberately buried canid were a wolf-dog according to anatomical and morphometric criteria, we would have some basis for hypothesizing such a treatment, but we cannot know which canid skull was buried where.

We can see that something is very strange about the treatment of the wolf-dogs at Předmostí beyond possible burial in a human cemetery area. One skull was buried with a bone of another animal wedged in its mouth (see Figure 13.1). The jaw was still attached to the cranium at the time that the bone—identified by Germonpré and colleagues as possibly from a mammoth—was placed in the mouth, so this cannot have occurred long after death. Some sort of special care was accorded to this wolf-dog at the time of death, but the precise meaning is elusive and haunting.[21]

In addition, an unusually high percentage of canid skulls from Předmostí (more than 40 percent) show premortem injuries to the jaw or face sufficient to break or remove teeth during life or to cause a facial fracture that showed signs of healing. Germonpré and her team showed that this incidence

Figure 13.1. One of the dogs from Předmostí had a large piece of bone, probably from a mammoth, inserted into its mouth shortly after death, perhaps as a mortuary ritual. This photo shows a three-quarter view of the skull and bone.

of trauma was significantly higher than that observed in a sample of thirty-five recent Eurasian wolves.

High frequencies of broken teeth in dire wolves in North America and other extinct carnivores prior to the arrival of humans have been taken as evidence of intense intraguild competition for food.[22] High tooth breakage in Předmostí might suggest the same: that competition among carnivores was fierce. This does not necessarily explain the unusually high percentage of facial injuries. The survival of injured dogs after facial injury might hint at care and feeding by modern humans. Conversely, the wolf-dogs at Předmostí might have been beaten by humans. There is no way to tell which explanation is correct, but wolf teeth do not show the same high incidence of breakage, and wolf skulls are not similarly broken.

A similar finding was reported by Robert Losey of the University of Alberta and colleagues, who studied 144 dogs and 400 wolves from northern Russian and archaeological sites in North America, Greenland, Kamchatka, and other far northern regions.[23] They found that loss of at least one tooth was overwhelmingly more common in dogs (54 percent) than in wolves (17 percent). Males and females were about equally likely to lose teeth. Dogs also lost significantly more teeth per individual than wolves and had a significantly higher frequency of tooth fracture than wolves. Ethnographic data indicate that dogs were provisioned, particularly with marine mammal blubber, organs, bones, and skin or frozen fish. However, provisioning was not particularly generous, and dogs may have competed among themselves for food. Wolves generally fed off of large or small terrestrial mammals, including hare, elk, moose, and sheep. Dogs also showed significantly more cranial injuries than wolves, particularly on the frontal bones. These injuries were probably associated with human disciplinary actions.

In addition, two canid teeth from Předmostí were modified by modern humans for wearing, probably as a pendant. The preparation of canine teeth of arctic foxes, red foxes, and wolves for wearing as ornaments is quite common in modern human sites dating to the Aurignacian or Gravettian. Randall White from New York University is an expert in prehistoric objects of personal adornment (jewelry). He argues that the teeth that are chosen for use as ornaments are deliberately selected and do not reflect the animals that dominate the faunal assemblages. He writes: "The animals whose teeth are worn [as ornaments] are not those whose meat is consumed. Phrased another way, the consumed fauna and the displayed fauna are almost mutually exclusive. This implies that the animals behind the parts transformed into ornaments are construed in terms that are largely of the collective symbolic imagination."[24] Though cut marks from bones at some sites suggest that wolves and foxes were sometimes eaten, the wolf and fox teeth were not chosen as personal adornment because they were common. In fact, they were uncommon. Reindeer or other herbivores far outnumber carnivores at many sites, and yet herbivore teeth were less often chosen as raw material in manufacturing objects of personal adornment. In fact, wolf and fox teeth might actually have been selected because they were generally rare or special, which apparently happened at Pavlov I. Teeth of wolves, foxes, or wolf-dogs may have been chosen as ornaments because they were symbolically powerful in some way. A simple—and perhaps slightly simple-minded—interpretation might be that people who modified and wore wolf or wolf-dog teeth were the People of the Wolf (or wolf-dog), who had a special relationship with and a unique understanding of wolves and wolf-dogs.

This idea may be a clue to the underlying reason that canids of any kind are rarely depicted in prehistoric cave art, such as paintings, engravings, carvings, and statues of clay, bone, or ivory. The oldest clearly identifiable and figurative art made by modern humans in Europe dates to about 32,000 years ago: roughly the same time as the earliest of the mammoth megasites. Overall and at any particular sites, images of carnivores as a group and of canids in particular are exceptionally uncommon. Paul Bahn, an independent scholar who is a noted expert in the field, once shared his thoughts with me on the paucity of canids in prehistoric art: "Probably [they are rare] for the same reason that human figures were also very rare—either it was taboo to depict them for some reason; or (far more likely) they were simply not relevant to what the artists were mostly doing."[25] Anne Pike-Tay of Vassar College, another specialist in the Upper Paleolithic, offered a slightly different twist: "The scarcity of artist depictions of carnivores parallels their scarcity in the fossil faunas of the Upper Paleolithic [of Western Europe]. What if dogs were put into the 'human family' category as an extension of the hunter and, like humans, warranted no (or very few) painted or engraved depictions?"[26]

Even at the mammoth megasites, some of which include exceptional numbers of canid bones from animals that were apparently skinned for their furs, neither wolves nor dogs nor arctic foxes are shown in the art objects. Ivory, bone, stone, and fired clay objects depict mammoths, horses, bison, bears, felids, water birds, lion, and Venus figurines. No dogs.

In marked contrast to the Aurignacian and Gravettian sites, Neanderthal sites contain few bones of foxes or wolves, and as yet, none of the canids from Neanderthal sites have been identified as a wolf-dog. Of the very few items in Neanderthal sites that could be considered art, none depict wolves, foxes, or any other type of canid. Whatever abilities modern humans used to capture and apparently domesticate wolves into wolf-dogs were either unknown to Neanderthals or beyond their capabilities. Both foxes and wolves were part of the Neanderthal ecosystem, but they apparently were not exploited systematically. Although domesticated wolf-dogs could in theory have been stolen from modern humans by Neanderthals, there is no evidence that they were. Unlike stone tools, dogs do not necessarily work for anyone who tries to use them.

Why didn't Neanderthals steal wolf-dogs or domesticate their own? A real possibility, and an important one, is based on the observation that there are no well-dated Neanderthal sites younger than about 40,000 years ago; all are older. Unless future finds show wolf-dogs in even earlier sites, Neanderthals were extinct by the time wolf-dogs appeared. The extinction of Neanderthals left wolves as the closest competitor with modern humans, who in turn challenged wolves for dominance by breeding and enlisting the help of wolf-dogs or wolves of their own.

Another serious possibility is that Neanderthals did not have or could not continue to transmit the requisite cognitive skills or knowledge to domesticate an animal or work with one. Kim Sterelny has suggested that, because Neanderthals were forced by climate change to break up into small groups, they would have been especially vulnerable to loss of knowledge because of random events and thus to extinction.[27] In a small, isolated group, the death of one individual might spell the loss of knowledge of how to make particular tools or of a hunting technique, or it simply might mean the loss of detailed information about the distribution of edible plants or the location of caves or of rich hunting grounds. Specialized knowledge about how to tame and domesticate a wolf could readily be lost.

• 14 •

WHEN IS A WOLF NOT A WOLF?

This question lies at the heart of any discussion of the domestication of wolves into dogs. As Darcy Morey has written, "To emphasize the behavioral factor, dogs make their living in a fundamentally different way from wolves. Their specific mode of living varies, but it is always in close association with people."[1]

In other words, a dog is a wolf that acts like a dog and relates to people.

What exactly has to change to make a wolf a dog? A major issue is the animal's interest in and connection to people. In wolves, there is a critical period of socialization that begins at about two weeks of age, when wolf pups are still blind and deaf, meaning that most of their sensory input is smell- or touch-based. Things encountered during this four-week period are investigated without fear and are regarded as familiar and safe for the rest of the animal's life. In dogs, this period occurs two weeks later in life, when the pups can see, hear, and smell. The delay in this development may have been crucial in the domestication of dogs, for it is obviously the period during which interaction with humans—or lack thereof—will have a lasting influence.[2]

There is as yet no simple way to measure the "doggyness" attribute; it is perhaps no more than an attitude, but it appears to be a genetically controlled attitude. A simple summary is that dogs exposed to humans during that critical period of socialization are intensely interested in people, and wolves are not. Though we recognize dogs in the fossil and archaeological record by their morphology and their genetics, it is their behavior and their relationship with humans that makes them dogs and not wolves.

Because communication with humans is so fundamental to being a dog, I will repeat here an interesting but speculative idea that I have put forth previously.[3] Humans have white scleras—the part of the eyeball that surrounds the colored iris—and this is unusual. In fact, in a survey of this trait, Hiromi Kobayashi and Shiro Kohshima of the Tokyo Institute of Technology showed that modern humans are unique among extant primates in having highly visible white scleras as well as eyelids that expose much of the sclera.[4] In other primates, the dark scleras, similarly colored skin, and concealing eyelids tend to mask the direction in which the animal is looking, according to the Japanese team. But in humans, the white scleras and open eyelids make the direction of a person's gaze highly visible from a distance, particularly if that glance is directed in a more or less horizontal direction. They suggest that the changes in the human eye are adaptations to enhance the effectiveness of the gaze signal. I hypothesize that this mutation may have become very common in modern humans about 50,000 to 45,000 years ago, around the time that they first invaded Eurasia. A highly visible direction of gaze could have been a big advantage in hunting cooperatively with wolfdogs.

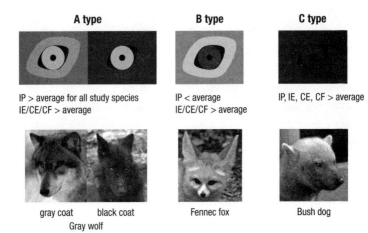

Figure 14.1. Sayoko Ueda and colleagues analyzed the faces of twenty-five species of living canids, measuring the color contrast among various features. The team divided all faces into three types. Type A faces showed strong contrast between the eye and the facial fur and between the pupil and the iris, making gaze direction particularly conspicuous. This occurred in group-hunting canids like wolves, foxes, jackals, coyotes, and dholes. Other patterns of coloration made the direction of gaze much less obvious (Type B) or actually concealed the direction of gaze (Type C). Types B and C were more common among solo hunting species.

Sayoko Ueda, working with Shiro Kohshima and other colleagues, has recently followed up on the observations on sclera color with a survey of color patterning surrounding the eyes in 320 individuals from twenty-five canid species.[5] Using photometric methods, the research team measured the contrast between pairs of facial segments—the iris to the pupil (IP), the iris to the eyelid (IE), the coat around the eyes to the eyelid (CE), and the coat around the eyes to the face (CF). They then divided the species into three types on the basis of the conspicuousness of the eye position within the face (see Figure 14.1).

Type A species have light-colored irises, dark pupils, and light-colored fur around the eyes. The color contrasts emphasize the direction of gaze and of the pupil itself to observers.

Type B species have light fur around the eyes, making their eye position clear, but dark irises that conceal the object on which they are focusing. Type C canids have camouflaged eyes—not outlined in lighter fur—and dark irises, which conceal both the direction of the gaze and the position of the pupil itself. The team found that color patterning varied significantly with hunting behavior or general sociality. Species that typically hunt or live in groups are more likely to have Type A faces, while those that hunt or live singly or in pairs have Type C faces.

Type A species have enhanced gaze signals—that is, their coloring makes their direction of gaze more obvious—and most belong to the wolf clade. In this study, the wolf clade includes gray wolves, coyotes, Ethiopian wolves, red wolves, dingoes, dholes, African wild dogs, golden jackals, side-striped jackals, and black-backed jackals. All group hunters among the wolflike canids have Type A faces except the dingo, which communicates by sound, and the African hunting dog, which signals conspecifics with its white-tipped tail. The dingo and the African hunting dog have Type B faces. No wolflike canids have Type C faces.

Because gray wolves can be black or gray in overall coat color, and they live in habitats ranging from subtropical (Mexico) to temperate and even high Arctic areas, the team examined whether the lightness of the iris varied with habitat and exposure to ultraviolet light. They found that the lightness of the iris in all gray wolves examined did not differ statistically with habitat.

Of the ten foxlike canids, five have Type B faces; three more social fox species—the arctic fox, the Corsican fox, and the Bengal fox—have Type A faces; and the remaining three have Type C faces. The Japanese team concluded that

Type A coloring probably evolved independently in these foxlike species and in wolves and correlated with hunting style.

Type C facial coloring was observed only in the bush dog in the third large group studied, a South American clade. Other canids in this group had Type B faces.

If Type A faces enhance the canids' abilities to communicate through gaze signals, as the researchers hypothesize, then the Type A species ought to gaze at others in their group for longer than the other facial types do. Focusing on gray wolves (Type A faces), fennec foxes (Type B faces), and bush dogs (Type C faces)—all group-living species—Ueda, Kohshima, and their colleagues studied the duration of gazing behavior in zoos among these animals. As they had predicted, the gray wolves (Type A faces) gazed toward each other for significantly longer periods of time than either the foxes (Type B faces) or the bush dogs (Type C faces).

Like humans, whose white sclera enhances their ability to communicate through gazes, the facial and eye coloring of wolves also seems to be an adaptation for gaze communication. And, the team notes, domestic dogs not only share the wolf's genetic ability to communicate through gazing, they also gaze at humans twice as long as wolves do on average— suggesting that duration of gaze may have been selected for during the domestication process. How fitting that both dogs and humans show an adaptation to improve visual communication of the direction of gaze!

The importance of following another individual's gaze has been explored by other researchers. In 2007, Michael Tomasello and colleagues at the Max Planck Institute of Evolutionary Anthropology developed the "cooperative eye hypothesis."[6] They suggested that cooperation among humans was facilitated by the ability to recognize where others were looking. Apes will follow the gaze of an experimenter less often than human infants will, they found. If the direction of gaze and the direction of the head conflicted, apes tended to follow head direction. In a humorous aside, the researchers noted that they tried their experiment with fourteen chimpanzees, four gorillas, four bonobos, and five orangutans—but dropped the results of tests on three chimps and all five orangutans because they "did not pay attention to the gaze cues sufficiently for their skills to be reliably assessed." Following human gaze was apparently not a high priority to the apes and offered no advantages.

The mutation causing white scleras is universal in humans, but it turns up occasionally in apes, too. In decades of observations at Gombe National Park in Tanzania, Jane Goodall and her team observed two chimps, probably brothers, who had white scleras.[7] A third, female, chimp developed white scleras as an adult. But the trait has not spread or reappeared in that population. The advantage of the white scleras must be related to something that ancient humans did commonly and chimps do not do or do rarely. One possibility is that the difference lies in hunting. Although chimps hunt small prey, often cooperatively, meat makes up less than 2 percent of their diet, whereas the earliest modern humans hunted much larger game that apparently provided a significant part of their diet. Silent communication among hunters would be clearly advantageous for working in human or canid groups or packs but would not matter so much to hunting apes.

Probably for the same reason, gray wolves are also skilled gaze readers, as the studies by Kohshima, Ueda, and colleagues show. However, the nonhuman species that seems to be most skilled at gaze reading is the domestic dog.

In experiments, a dog will follow the gaze of a videotaped human if the human on the tape first attracts the dog's attention by speaking to it and looking at it, according to results published by Ernő Téglás, of the Central European University in Budapest, and his colleagues. Indeed, dogs perform as well as human infants at following the gaze of a speaker in tests in which the speaker's head is held still.

Friederike Range and Zsófia Virányi have shown that socialized, hand-reared wolves will reposition themselves to follow a gaze signal from a human handler when the subject of the human's gaze cannot be seen by the wolf because of an intervening visual barrier.[8] This barrier test is generally considered to be a more complex or sophisticated demonstration of gaze following than simply looking in the direction of the handler. It has been suggested that this type of gaze following may be most useful in species with complex cooperative and competitive interactions, like wolves.

Ádám Miklósi of Eötvös Loránd University in Budapest and his team tested dogs and wolves and found that dogs were far more attentive to human faces than were wolves, even socialized wolves.[9] Although wolves excel at some gaze-following tasks, perhaps suggesting a preadaptation for communicating with humans, dogs tend to look at human faces for cues in novel situations and wolves do not. Miklósi's team believes this major behavioral difference is the result of selective breeding that transformed wolves into dogs during domestication.

Dogs look to humans for guidance; dogs emulate human activities in solving problems; dogs follow human gestures like pointing, body position, and the direction of the human vision or gaze. Dogs even turn their own gaze toward humans and initiate eye contact as a communicative signal. Wild wolves do not. In fact, wild wolves may interpret prolonged staring by a human as a threat gesture.

My idea is that white scleras became universal among humans because it enabled them to communicate better not only with each other but also with the wolf-dogs with which they lived and hunted. Once these animals could read a human gaze signal, they would have been even more useful as hunting partners than canids that could simply run fast and had a better sense of smell than modern humans. The wolves ancestral to dogs apparently evolved to have better gaze following than other canids, and dogs themselves have adapted still further by behavioral changes that involve paying special attention to humans.

I have been unable to find research identifying the genetic basis of white scleras. Once this fact is known, then inspection of the genomes of Neanderthals and modern humans should reveal the presence or absence of white scleras. But if the white sclera mutation occurred more often among modern humans—perhaps by chance—this feature could have enhanced communication between canids and humans, promoting domestication. A lack of white scleras among Neanderthals, if shown to have existed, might have hindered their ability to work with wolf-dogs and other humans.

Humans today love to look into their dogs' eyes to "read" their emotions; it is part of bonding. Dogs apparently feel the same and will gaze into a human's eyes in an attempt to communicate. The mutation that originally produced the white sclera trait in humans can be viewed as a key aid to reciprocal communication with dogs and may have been instrumental in the survival of our species over Neanderthals.

What we see in the fossil and archaeological record is that by about 36,000 (calibrated) years ago, there was a distinctive group of large canids that showed up unusually often in sites made by humans, particularly where those humans hunted mammoths and other large animals with a success rate never before achieved. Some factor—quite probably the presence of those wolf-dogs—enabled humans to retain control of these carcasses for weeks or perhaps months. The faunal remains left at these sites were different from those at earlier sites, not only in terms of the dramatic increase in mammoth carcasses but also in the abundance of wild wolves, along with arctic foxes, red foxes, and hares, which became especially preferred prey for the first time. Teeth of these species were favored items for modification for use as pendants or other items of adornment, suggesting modern humans had a new view of these species. Sites are larger and more complex with distinct and identifiable activity areas and longer periods of occupation. The density of those sites increases, signaling a steady expansion of the space, housing, food, and tools provided for more people through time.

From about the time of modern human arrival, we see a contraction of Neanderthal territory, fewer Neanderthal sites, and smaller sites containing a lower density of tools and prey remains. Genetic diversity among Neanderthals lessened as well. There is also a population diminution in terms of the numbers of Eastern cave bears, lesser scimitar cats, cave lions, and cave hyenas—all actual or potential competitors with modern human hunters. The genetic variability of cave bear populations diminishes and the bears that continue to live in areas also inhabited by modern humans shift toward a more vegetarian diet, perhaps to lessen competition with modern humans. The temperate woodlands and forests shrink in ab-

solute size, and their location is restricted to farther and farther south, as the mammoth steppe habitats and their faunas
expand.

If modern humans acted as an invasive predatory species
when they entered the Eurasia ecosystem about 50,000 to
45,000 years ago, many peculiarities in the fossil record are
explained. What made modern humans especially powerful
apex predators was their alliance with another apex predator.
No other predator has done this to such an extent, though
sometimes predators use the calls or presence of another species to locate a fresh carcass. Hyenas watch vultures, for example, and hyena calls may alert lions. But none has undertaken the transformation that enables humans and former
wolves to live together, work together, and develop intense
forms of interspecies communication. If additional evidence
continues to support the hypothesis that modern humans reinforced their role as apex predator by allying themselves with
wolf-dogs, nearly all of the remaining questions about modern
human survival and Neanderthal extinction can be answered.
In Figure 14.2, artist Dan Burr shows an image of what the
early human-dog alliance might have looked like.

Though climate change undoubtedly accompanied the extinction of the Neanderthals, climate change is an unsatisfactory candidate as the primary driving force. It fails to answer the key question of why Neanderthals did not go extinct
during previous cold phases as severe and as long-lasting as
the ones during MIS 3. With redating of various sites, there
is no longer an indication that Neanderthals retreated southward to milder climates during MIS 3, as would be expected
if climate change were the driving force in Neanderthal extinction. What was different about the time during which Neanderthals went extinct was the added pressure of an invasive

Figure 14.2. The artist Dan Burr created this image of early modern humans hunting mammoth with wolf-dogs. The landscape is reminiscent of that near the sites of Dolní Věstonice and Pavlov in the Czech Republic.

apex predator that competed strongly with Neanderthals and then paired up with a third apex predator. The entire predator guild was threatened and challenged by the appearance of modern humans, not just Neanderthals, and many predators went locally or globally extinct.

The unprecedented alliance of modern humans with another top predator (wolf-dogs or a kind of wolf) may have been the final stress that pushed Neanderthals and many other species down the slippery slope toward extinction, if Neanderthals were still extant when wolf-dogs first appeared. Certainly the colder, drier climates wreaked important effects on the habitats used by Neanderthals and most other predators—

though gray wolves themselves are today more often found in cold habitats. Probably the unpredictable climate of MIS 3 worked to reduce the population size and geographic distribution of Neanderthals and other archaic species. For example, studies of global climate changes through time reveal that continents that experienced colder, harsher climates during the Last Glacial Maximum experienced more extinctions.[10] But climate change in and of itself was not sufficient to force Neanderthals into extinction. Climate change combined with increased intraguild competition spelled trouble, particularly for Neanderthals whose metabolic needs were so high during colder spells. What we see is a classic trophic cascade caused by the appearance of modern humans that influenced all the other species in the ecosystem.

WHAT HAPPENED AND WHY

Looking back over the ideas and evidence I have synthesized here, there is a remarkably consistent story—but one that is still full of holes and unknowns.

Starting at perhaps 42,000 years ago or slightly earlier, there was a contraction of Neanderthal territory and a slowing down or an actual decline in Neanderthal sites in terms of frequency, size, density of tools, and amount of prey taken. Genomic studies indicate that there was a genetic bottleneck in Neanderthal diversity at about this time, which almost certainly reflects a much-diminished population that had broken up into smaller groups.[1]

There is also evidence of a genetic bottleneck in the mtDNA of Eastern cave bears, in cave lions, and in cave hyenas after roughly 40,000 years ago, and some of the most striking declines came at about 32,000 years ago. The lesser scimitar cat, *Homotherium latidens,* apparently survived until about 28,000 years ago before going extinct, but this species is known from only two specimens younger than 200,000 years ago. This extinction date thus needs confirmation because there is such a long chronological gap between the specimens at 28,000 years ago and the previous fossils. The lesser scimitar cat may have been extinct well before the arrival of modern humans, or as paleontologists who reported the most recent lesser scimitar cat have suggested, some of the isolated teeth of this species may have been mistaken for those of cave lions.[2] The timing of dhole extinction is uncertain, but may also fall between 40,000 and 35,000 years ago.[3]

These loose chronological coincidences in extinction prompt me to look for a common cause, or two, for these patterns in hominins and predators. All of the carnivorous species actually or potentially competed with modern human hunters for the same prey and in many cases for the most desirable caves for habitation or denning in Eurasia.[4] Though the genetic diversity of cave bear populations waned—as did that of Neanderthals themselves—the bears that continued to live in areas also inhabited by modern humans adopted a more vegetarian diet. There is no sign that Neanderthals responded to the heightened competition by similarly changing their diet. Late and early, in forests or more open areas, Neanderthals seemed to rely heavily on the same prey species. The temperate woodlands and forests were shrinking in absolute size, and some of the competing carnivores were restricted to areas that were farther and farther south, as the northern mammoth steppe habitats and their faunas expanded. Wolves were one of the exceptions to this rule. To summarize bluntly, there was a massive crash in the carnivore guild starting around 40,000 years ago and culminating in extinctions or extirpations—the last gasps of several species— by around 32,000 years ago.

All of these species had already survived climate change episodes as dramatic as those in MIS 3, and all of them had already survived increased cold and changing vegetation.

Genetic studies suggest that the populations of some of these species became fragmented and lived in isolated refugia, only to recolonize larger ranges when conditions improved.

Climate change undoubtedly accompanied and fostered the extinction of the Neanderthals and the other now-extinct predators of the Pleistocene, but climate change is an unsatisfactory candidate as the primary driving force because it fails to answer some key questions. Climate change does not explain why Neanderthals did not go extinct during previous cold phases that were as severe and long-lasting as the ones during MIS 3. If climate change drove Neanderthals southward—and there is now no strong evidence that it did— why couldn't they survive and recolonize the more northerly regions when the climate improved? Climate change does not identify any advance or new technical ability possessed by humans that was not possessed by Neanderthals, which would enable modern humans to survive when Neanderthals could not. Climate change does not account for the survival of one hominin species over the other and does not have any obvious bearing on the extraordinary changes that occurred in archaeological sites made by modern humans between 40,000 and 30,000 years ago.

I argue that the technical advance that made humans so irresistible and so invasive—from 50,000 years ago until today—was in part their ability to form this unprecedented alliance with another species that we call domestication. We turned wolves into dogs and much later, mouflon into goats, wild aurochs into cattle, wild cats into house cats, and horses into a rapid transit system. We created for ourselves an ability to borrow the traits of other species and use them to enhance our own survival in almost every habitat on the planet.

I suggest the combination of climate change and the arrival of modern humans with new abilities acted together to cause Neanderthal extinctions. Complex modeling carried out by Graham Prescott and colleagues of the roles of climate change, human actions, or both to explain the megafaunal extinctions on five continental landmasses led to the same conclusion.[5] They found that an excellent climatic predictor of extinction was the most rapid rate of temperature decline during a period, rather than the mean temperature per se. In other words, it was the rapidity of change, not the magnitude of change, that mattered. Biological adaption is a slow means of coping with climate change, whereas behavioral flexibility and innovation work rapidly. Adding human arrival and competition on top of rapid temperature decline consistently improved the fit of the models Prescott and colleagues tested against reality. It seems likely that the unpredictable climate of MIS 3 reduced the population size and geographic distribution of Neanderthals and other archaic species, as studies of global climate changes through time reveal that continents that experienced colder, harsher climates during the Last Glacial Maximum also experienced more extinctions.[6] But climate change in and of itself was not sufficient to force Neanderthals into extinction. If so, they would have died out many millennia earlier.

Given the tremendous stability in Neanderthal diets and tools over hundreds of thousands of years prior to their extinction, it appears that they were slow to innovate and slow to change their ways in their world. The unparalleled alliance of modern humans with another top predator (wolf-dogs) may have been the final strategy that made survival of Neanderthals and many other predatory species impossible. Climate change combined with a new and heightened form of intra-guild competition spelled serious trouble. What we can see in the archaeological and paleontological record looks like a classic trophic cascade caused by the appearance of an apex predator. This event would have been particularly instrumental in altering the interrelationships among all the other species in the ecosystem because of climate change.

The strengths of the answer I propose here to the question "why did Neanderthals go extinct while we survived?" are several.

First, my scenario factors in the importance of intraguild competition caused by an invasive new apex predator and also the importance of working with wolf-dogs as a means of adapting to climate change. This approach places hominin evolution within the context of mammalian evolution and ecological theory and practice. By using the insights afforded by invasion biology to guide my interpretation of the evidence, I am looking at our ancestors as part of nature and a vital part of the ecological community, not as an utterly separate crea-ture that was exempt from ecological and biological rules. Hu-mans are animals; we are mammals; we have the same basic needs to eat, survive, reproduce, and raise our offspring as any other mammalian species. We have developed some extraor-dinary ways of fulfilling those needs, by using language, tools in the broadest sense, and alliances both within and without our species. We may have enhanced our physical abilities ge-netically, by evolving new means of making gaze communi-cation (and possible other forms of communication) more ef-ficient. Improved communication may have improved our success in obtaining the resources we needed for survival. Nonetheless, we are not freed from the sometimes harsh realities of life on earth. If we outrun our resources, we will be faced with the same death knell that many before us have also heard. We are a powerful part of many terrestrial ecosystems, and we need to understand our role and our responsibilities to our fellow species. I think this is a vitally important perspective.

Second, my scenario may explain the Neanderthal extinction, the crash in the predator guild, and the abrupt appearance of the Gravettian mammoth megasites. If working with wolf-dogs began even a few millennia earlier than its earliest known appearance (presently about 36,000 calibrated years ago), this behavioral change may have been an essential component of early modern humans' adaptability in response to competition and climate change. I expect that whenever wolf-to-dog domestication began, the process was not very efficient at first and involved a long period of trial and error. Domesticating the first animal was one of the huge steps our species has taken during our evolution, a step on a par with the initial invention of tools. Domestication of other species has had an enormous impact on the opportunities and options open to human beings. We did not have to develop profound speed in running to catch elusive prey, for example; all we had to do was form an alliance with another species that was faster than we were. We did not have to evolve a keener sense of smell or eyesight because we could "borrow" these traits from wolves. Domesticates have made possible the invasion and growth in human populations in many different habitats and on many different continents.

Third, I have been able to integrate recent findings and revelations from genomic and genetic studies, some of which seem to resolve problems and some of which pose more questions than answers. Issues like the admixture of Neanderthals and modern humans, once thought to have been definitively ruled out, are now known to have occurred, albeit infrequently. The hints that the genes implicated in male fertility were strongly selected against in hybrids between Neanderthals and modern humans and may explain why interbreeding was not a successful long-term strategy for Neanderthals. With the growing information and superb technical skills of genomic researchers, we are now able to go beyond the bones and stone of the past—complex as those are and rich as they are as indicators of behavior—to add into the mix the basic biology of the species we examine.

Fourth, by evaluating the impact of climate change in addition to our alliance with another predator, I have generated a series of predictions based on my hypothesis and have been able to present some evidence to evaluate the likelihood of their being accurate. New evidence about Neanderthal and early modern human lifestyles in this crucial period increases every day. We acquire new information, make new comparisons, obtain new finds, and develop new analyses that shift the picture, massively or slightly.

Finally, the immense advance in dating techniques has been crucial because it has given us a stable chronological framework within which various hypotheses can be evaluated and tested. The redating of materials has been an essential component in reaching a new synthesis of information and in testing hypotheses new and old.

As paleoanthropology progresses, some new material is bound to surprise or shock us; I think both Neanderthals and ancient modern humans were far more able and sophisticated than they are usually given credit for being. Certainly, the process of research and discovery about our past is ongoing. I hope in a few years to look back on this book and wonder how I (and everyone else) missed so much.

There are weaknesses in the synthesis I present here: unanswered questions, unmeasured parameters, holes in the evidence where we have nothing at all to inform us yet. Dates on many sites are still unreliable. But the excitement in the air and the vibrancy of the discussion, both at a popular and a scientific level, make me think that all of the diverse evidence is starting to come together. Once it does, we may understand our role in the world's ecosystems more precisely, more objectively. This perspective is a necessary first step toward balancing the demands of humans and of other organisms on this earth more wisely. We may be more successful at preserving biological diversity and truly wild places as we see more clearly the factors that diminish them. We may be able to envision more accurately humans' place in nature, for the good of all.

I think it is time we recognized ourselves for what we are: the invaders. If one day we can meet the enemy of the earth and he is not us, it will be a triumph. But a big change in our behavior must happen first.

NOTES

1. And He Is Us

1. W. Kelly, "Pogo," http://www.igopogo.com/we_have_met.htm.
2. P. Shipman, *The Animal Connection* (New York: Norton, 2012).
3. T. Flannery and D. Burney, "Fifty Millennia of Catastrophic Extinctions after Human Contact," *Trends in Ecology and Evolution* 20 (2005): 395–401.
4. Ibid., 395.
5. P. Martin and H. Wright, *Pleistocene Extinctions* (New Haven, CT: Yale University Press, 1967).
6. J. Estes, J. Terborgh, J. Brashares et al., "Trophic Downgrading of Planet Earth," *Science* 333 (2011): 301–306.
7. I. Davidson, "Peopling the Last New Worlds: The First Colonisation of Sahul and the Americas," *Quaternary International* 285 (2013): 1–29; J. Field, S. Wroe, C. Trueman et al., "Looking for the Archaeological Signature in Australian Megafaunal Extinctions," *Quaternary International* 285 (2013): 76–88; S. Wroe, J. Field, and D. Grayson, "Megafaunal Extinctions: Climate, Humans and Assumptions," *Trends in Ecology and Evolution* 21 (2006): 51–52.
8. J. Ray, K. Redford, R. Stenecke et al., "An Ecological Context for the Role of Large Carnivores in Conserving Biodiversity," in *Large Carnivores and the Conservation of Biodiversity*, ed. J. Ray, K. Redford, R. Steneck, and J. Berger (Washington, DC: Island Press, 2005), 15.
9. P. Hortolà and B. Martínez-Navarro, "The Quaternary Megafaunal Extinction and the Fate of Neanderthals: An Integrative Working Hypothesis," *Quaternary International* 295 (2013): 69–72.

2. Here We Come, Ready or Not

1. Executive Order 13112 of February 3, 1999, *Invasive Species Federal Register* 64, no. 25, February 8, 1999, Presidential Documents 6183.
2. Ibid.
3. G. J. Vermeij, "Invasion as Expectation: A Historical Fact of Life," in *Species Invasions: Insights into Ecology, Evolution, and Biogeography*, ed. D. Sax, J. Stachowicz, and S. Gaines (Sunderland, MA: Sinauer Associates, 2005), 315–339.
4. L. Traill, J. Corey, C. Bradshaw et al., "Minimum Viable Population Size: A Meta-Analysis of 30 Years of Published Estimates," *Biological Conservation* 1339 (2007): 159–166.
5. J. O'Connell and J. Allen, "The Restaurant at the End of the Universe: Modelling the Colonisation of Sahul," *Australian Archaeology* 74 (2012): 5–31.
6. Davidson, "Peopling the Last New Worlds."
7. A. Williams, "A New Population Curve for Prehistoric Australia," *Proceedings of the Royal Society B* 280 (2013): 20130486, http://dx.doi.org/10.1098/rspb.2013.0486; O'Connell and Allen, "Restaurant at the End of the Universe."
8. T. Watson, "Who Were the First Australians, and How Many Were There?" *Science Now,* April 23, 2013, quoting Alan Williams.
9. J. Lockwood, M. Hoopes, and M. Marchetti, *Invasion Ecology* (Malden, MA: Blackwell, 2007); T. Blackburn, P. Pyšek, S. Bacher et al., "A Proposed Unified Framework for Biological Invasions," *Trends in Ecology and Evolution* 26 (2011): 333–339, doi:10.1016/j.tree.2011.03.023; P. Hulme, S. Bacher, M. Kenis et al., "Grasping at the Routes of Biological Invasions: A Framework for Integrating Pathways into Policy," *Journal of Applied Ecology* 45 (2008): 403–414.
10. R. Leakey and R. Lewin, *The Sixth Extinction* (New York: Anchor Books, 1996); E. Kolbert, *The Sixth Extinction: An Unnatural History* (New York: Henry Holt, 2014).
11. D. Wilcove, D. Rothstein, J. Dubow et al., "Quantifying Threats to Imperiled Species in the United States," *BioScience* 48 (1998): 607–615; O. Sala, F. S. Chapin III, J. Armesto et al., "Global Biodiversity Scenarios for the Year 2100," *Science* 287 (2000): 1770–1774.
12. M. Clavero and E. García-Berthou, "Invasive Species Are a Leading Cause of Animal Extinctions," *Trends in Ecology and Evolution* 20 (2005): 110.

13. C. Elton, *The Ecology of Invasions by Animals and Plants* (London: Methuen, 1958).

14. M. Krings, H. Geisert, R. Schmitz et al., "DNA Sequence of the Mitochondrial Hypervariable Region II from the Neandertal Type Specimen," *Proceedings of the National Academy of Sciences USA* 96 (1999): 5581.

15. M. Krings, A. Stone, R. Schmitz et al., "Neandertal DNA Sequences and the Origin of Modern Humans," *Cell* 90 (1997): 19–30; R. Green, J. Krause, S. Ptak et al., "Analysis of One Million Base Pairs of Neanderthal DNA," *Nature* 444 (2006): 330–336; J. Noonan, "Neanderthal Genomics and the Evolution of Modern Humans," *Genome Research* 20 (2010): 547–553; J. Noonan, G. Coop, S. Kudaravalli et al., "Sequencing and Analysis of Neanderthal Genomic DNA," *Science* 314 (2006): 1113–1118; R. Green, J. Krause, A. Briggs et al., "A Draft Sequence of the Neandertal Genome," *Science* 328 (2010): 710–722, doi:10.1126/science.1188021; K. Prüfer, F. Racimo, N. Patterson et al., "The Complete Genome Sequence of a Neanderthal from the Altai Mountains," *Nature* 505 (2014): 43–49, doi:10.1038/nature12886.

16. "Neanderthal Genome Yields Insights into Human Evolution and Evidence of Interbreeding with Modern Humans," *Science News,* May 6, 2010.

17. Ibid.

18. A. Eriksson and A. Manica, "Effect of Ancient Population Structure on the Degree of Polymorphism Shared between Modern Human Populations and Ancient Hominins," *Proceedings of the National Academy of Sciences USA* 109 (2012): 13965–13960.

19. K. Prüfer, F. Racimo, N. Patterson et al., "The Complete Genome Sequence of a Neanderthal from the Altai Mountains," *Nature* 505 no. 7481 (2014): 43–49, doi: 10.1038/nature12886.

20. Sarah Tischkoff, personal communication to author, Feb. 5, 2014.

21. B. Vernot and J. M. Akey, "Resurrecting Surviving Neandertal Lineages from Modern Human Genomes," *Science Express*, January 29, 2014, 1–4, doi 2014/ 0.1126/science.1245938.

3. Time Is of the Essence

1. A. Froehle and S. Churchill, "Energetic Competition between Neandertals and Anatomically Modern Humans," *Paleoanthropology* (2009): 96–116; C. Ruff, E. Trinkaus, and T. Holliday, "Body Mass and Encephalization in Pleistocene Homo," *Nature* 387 (1997):

173–176; C. Ruff, personal communication to the author, July 25, 2012.

2. O. Bar-Yosef and J. G. Bordes, "Who Were the Makers of the Châtelperronian Culture?" *Journal of Human Evolution* 59 (2010): 586–593; T. Higham, R. Jacobi, M. Julien et al., "Chronology of the Grotte du Renne (France) and Implications for the Context of Ornaments and Human Remains within the Châtelperronian," *Proceedings of the National Academy of Sciences USA* 107 (2010): 20234–20239; B. Gravina, P. Mellars, and C. Bronk Ramsey, "Radiocarbon Dating of Interstratified Neanderthal and Early Modern Human Occupations at the Châtelperronian Typesite," *Nature* 438 (2005): 51–56; P. Mellars, B. Gravina, and C. Bronk Ramsey, "Confirmation of Neanderthal/Modern Human Interstratification at the Châtelperronian Type-Site," *Proceedings of the National Academy of Sciences USA* 104 (2006): 3657–3662; J. Zilhão, F. d'Errico, J.-G. Bordes et al., "Analysis of Aurignacian Interstratification at the Châtelperronian Type Site and Implications for the Behavioral Modernity of Neandertals," *Proceedings of the National Academy of Sciences USA* 103 (2006): 12643–12648.

3. F. Caron, F. d'Errico, P. Del Moral et al., "The Reality of Neandertal Symbolic Behavior at the Grotte du Renne, Arcy-sur-Cure, France," *PLoS ONE* 6 (2011): e21545, doi:10.1371/journal.pone.0021545.

4. R. Wood, C. Barroso-Ruíz, M. Caparrós et al., "Radiocarbon Dating Casts Doubt on the Chronology of the Middle to Upper Palaeolithic Transition in Southern Iberia," *Proceedings of the National Academy of Sciences USA* 110 (2013): 2781–2786, doi:10.1073/Proceedings of the National Academy of Sciences USA.1207656110.

5. T. Higham, K. Douka, R. Wood et al., "The Timing and Spatio-Temporal Patterning of Neanderthal Disappearance," *Nature* 512 (2014): 306–309, doi:10.1038/nature13621. Supplementary Methods.

6. R. Wood, K. Douka, P. Boscato et al., "Testing the ABOx-SC Method: Dating Known-Age Charcoals Associated with the Campanian Ignimbrite," *Quaternary Geochronology* 9 (2012): 16–26.

7. Wood et al., "Radiocarbon Dating Casts Doubt," 4.

8. E. Callaway, "Date with History," *Nature* 485 (2013): 27–29; R. Pinhasa, T. Higham, L. Golovanova et al., "Revised Age of Late Neanderthal Occupation and the End of the Middle Paleolithic in

the Northern Caucasus," *Proceedings of the National Academy of Sciences USA* 108 (2011): 8616.

9. J. Amos, "Last-Stand Neanderthals Queried," BBC News, February 5, 2013, http://www.bbc.co.uk/news/science-environment-21330194.

10. T. Higham, personal communication to author, June 28, 2013.

11. L. Golovanova, B. Doronichev, N. Cleghorn et al., "Significance of Ecological Factors in the Middle to Upper Paleolithic Transition," *Current Anthropology* 51 (2010): 655–691.

12. Pinhasa et al., "Revised Age of Late Neanderthal Occupation."

13. T. Higham, L. Basell, R. Jacobi, R. Wood et al., "Testing Models for the Beginnings of the Aurignacian and the Advent of Figurative Art and Music: The Radiocarbon Chronology of Geissenklösterle," *Journal of Human Evolution* 62 (2012): 665.

14. T. Higham, T. Compton, C. Stringer et al., "The Earliest Evidence for Anatomically Modern Humans in Northwestern Europe," *Nature* 479 (2011): 521–524, doi:10.1038/nature10484.

15. Higham et al., "Timing and Spatio-Temporal Patterning."

4. Who Wins in an Invasion?

1. P. Savoleinen, T. Leitner, A. Wilton et al., "A Detailed Picture of the Origin of the Australian Dingo, Obtained from the Study of Mitochondrial DNA," *Proceedings of the National Academy of Sciences USA* 101 (2004): 12387–12390.

2. Shipman, *The Animal Connection.*

3. J. Zalasiewicz, M. Williams, A. Smith et al., "Are We Now Living in the Anthropocene?" *GSA Today* 18, no. 2 (2008): 4–8, doi:10.1130/GSAT01802A.1; J. Zalasiewicz, M. Williams, W. Steffen et al.,"The New World of the Anthropocene," *Environmental Science and Technology* 44 (2010): 2228–2231; P. Crutzen and E. Stoermer, "The 'Anthropocene,'" *Global Change Newsletter* 41 (2000): 17–18.

4. U. Müller, J. Pross, P. Tzedakis et al., "The Role of Climate in the Spread of Modern Humans into Europe," *Quaternary Sciences Review* 30 (2011): 273–279. National Climatic Data Center, "A Paleo Perspective on Abrupt Climate Change," http://www.ncdc.noaa.gov /paleo/abrupt/data3.html, gives a good overview.

5. J. Lowe, N. Barton, S. Blockley et al., "Volcanic Ash Layers Illuminate the Resilience of Neanderthals and Early Modern Humans to Natural Hazards," *Proceedings of the National Academy of Sciences USA* 109 (2012): 13532–13537, doi: 10.1073/pnas.1204579109;

J.-J. Hublin, "The Earliest Modern Human Colonization of Europe," *Proceedings of the National Academy of Sciences USA* 109 (2012): 13471–13472.

6. Lowe et al., "Volcanic Ash Layers," 13532.

7. Ibid., 13536.

8. B. Arensburg, "Human Remains from Geula Cave, Haifa," *Bulletins et mémoires de la Société d'Anthropologie de Paris* 14 (2002): 141–148.

9. J. Shea, "The Archaeology of an Illusion: The Middle Paleolithic Transition in the Levant," in *The Lower and Middle Palaeolithic in the Middle East and Neighboring Regions*, eds. J.-M. Le Tensorer, R. Jagher, and M. Otte (Basel Symposium, Eraul, Liège, 2008), 169–182.

10. K. Brown, C. Marean, Z. Jacobs et al., "An Early and Enduring Advanced Technology Originating 71,000 Years Ago in South Africa," *Nature* 491 (2012): 590–593.

11. Shea, "Archaeology of an Illusion," 177.

12. F. Demeter, L. Shackelford, A.-M. Bacon et al., "Anatomically Modern Human in Southeast Asia (Laos) by 46 ka," *Proceedings of the National Academy of Sciences USA* 109 (2012): 14375–14380.

13. A. Pierret, V. Zeitoun, and H. Forestier, "Irreconcilable Differences between Stratigraphy and Direct Dating Cast Doubts upon the Status of Tam Pa Ling Fossil," *Proceedings of the National Academy of Sciences USA* 109 (2012): E3523, www.pnas.org/cgi/doi/10.1073/pnas.1216774109; but see also F. Demetera, L. Shackleford, K. Westaway et al., "Reply to Pierret et al.: Stratigraphic and Dating Consistency Reinforces the Status of Tam Pa Ling Fossil," *Proceedings of the National Academy of Sciences USA* 109 (2012): E3524–E3525.

14. C. Bae, W. Wang, J. Zhao et al., "Modern Human Teeth from Late Pleistocene Luna Cave (Guangxi, China)," *Quaternary International* (in press), Corrected Proof, Available online July 28, 2014.

15. Higham et al., "Earliest Evidence for Anatomically Modern Humans"; S. Benazzi, K. Douka, C. Fornai et al., "Early Dispersal of Modern Humans in Europe and Implications for Neanderthal Behavior," *Nature* 479 (2011): 525–528; T. Higham, C. Stringer, and K. Douka, "Dating Europe's Oldest Modern Humans," *British Archaeology*, January–February 2012, 24–29.

16. E. Trinkaus, O. Moldovan, S. Milota et al., "An Early Modern Human from the Peştera cu Oase, Romania," *Proceedings of the*

National Academy of Sciences USA 100 (2003): 11231–11236; Higham et al., "Dating Europe's Oldest."

5. How Do You Know What You Think You Know?

1. C. Finlayson, *Neanderthals and Modern Humans: An Ecological and Evolutionary Perspective* (Cambridge: Cambridge University Press, 2004).
2. Ibid., 153.
3. Ibid., 154.
4. A. Cahill, M. Aiello-Lammens, M. Fisher-Reid et al., "How Does Climate Change Cause Extinction?" *Proceedings of the Royal Society B* 280 (2013): 2–3.
5. Finlayson, *Neanderthals and Modern Humans*, 153.
6. See discussions in E. Trinkaus and P. Shipman, *The Neanderthals* (New York: Knopf, 1992).
7. C. Marean, personal communication to author, January 25, 2014.
8. G. F. Gause, *The Struggle for Existence* (Baltimore: Williams & Wilkins, 1934).
9. G. Hardin, "The Competitive Exclusion Principle," *Science* 131 (1960): 1292–1297.
10. J. Terborgh, R. Holt, and J. Estes, "Trophic Cascades: What They Are, How They Work, and Why They Alter," in *Trophic Cascades: Predators, Prey, and the Changing Dynamics of Nature*, eds. J. Terborgh and J. Estes (Washington, DC: Island Press, 2010), 7.
11. See R. MacArthur and E. O. Wilson, *The Theory of Island Biogeography* (1967; repr., Princeton, NJ: Princeton University Press, 2001).
12. A. Portmann, *A Zoologist Looks at Humankind* (New York: Columbia University Press, 1990); T. Smith, P. Tafforeau, D. Reid et al., "Dental Evidence for Ontogenetic Differences between Modern Humans and Neanderthals," *Proceedings of the National Academy of Sciences USA* 107 (2010): 20923–20928.

6. What's for Dinner?

1. G. Roemer, M. Gompper, and V. Van Valkenburgh, "The Ecological Role of the Mammalian Mesocarnivore," *BioScience* 59 (2009): 165–173.
2. J. Grinnell, "The Origin and Distribution of the Chestnut-Backed Chickadee," *The Auk* (American Ornithologists' Union) 21 (1904): 364.

3. B. Van Valkenburgh, "Tracking Ecology over Geological Time: Evolution within Guilds of Vertebrates," *Trends in Evolution and Ecology* 10 (1995): 71–76; W. Ripple and B. Van Valkenburgh, "Linking Top-Down Forces to the Pleistocene Megafaunal Extinctions," *BioScience* 60 (2010): 516–526.

4. M. Patou-Mathis, "Neanderthal Subsistence Behaviours in Europe," *International Journal of Osteoarchaeology* 10 (2000): 379–395; D. Grayson and F. Delpech, "The Large Mammals of Roc de Combe (Lot, France): The Châtelperronian and Aurignacian," *Journal of Anthropological Archaeology* 27 (2008): 338–362; D. Grayson and F. Delpech, "Changing Diet Breadth in the Early Upper Paleolithic of Southwestern France," *Journal of Archaeological Science* 25 (1998): 1119–1130; D. Grayson and F. Delpech, "Specialized Early Upper Paleolithic Hunters in Southwestern France?" *Journal of Archaeological Science* 29 (2002): 1439–1449; D. Grayson and F. Delpech, "Ungulates and the Middle-to-Upper Paleolithic Transition at Grotte XVI (Dordogne, France)," *Journal of Archaeological Science* 30 (2003): 633–640; D. Grayson and F. Delpech, "Pleistocene Reindeer and Global Warming," *Conservation Biology* 19 (2005): 557–562; D. Grayson and F. Delpech, "Was There Increasing Dietary Specialization across the Middle-to-Upper Paleolithic Transition in France?" in *When Neanderthals and Modern Humans Met,* ed. N. Conard (Tübingen: Kerns Verlag, 2006), 377–417; D. Grayson, F. Delpech, J.-Ph. Rigaud et al., "Explaining the Development of Dietary Dominance by a Single Ungulate Taxon at Grotte XVI, Dordogne, France," *Journal of Archaeological Science* 28 (2001): 115–125; M. Stiner, *Honor among Thieves: A Zooarchaeological Study of Neandertal Ecology* (Princeton, NJ: Princeton University Press, 1994).

5. C. Stringer, J. C. Finlayson, R. Barton et al., "Neanderthal Exploitation of Marine Mammals in Gibraltar," *Proceedings of the National Academy of Sciences USA* 105 (2008): 14319–14324.

6. O. Bar-Yosef, "Eat What Is There: Hunting and Gathering in the World of Neanderthals and Their Neighbors," *International Journal of Osteoarchaeology* 14 (2004): 333–342, doi:10.1002/oa.765.

7. M. Richards and E. Trinkaus, "Isotopic Evidence for the Diets of European Neanderthals and Early Modern Humans," *Proceedings of the National Academy of Sciences USA* 106 (2009): 16034–16039.

8. M. Richards, P. Pettitt, M. Stiner et al., "Stable Isotope Evidence for Increasing Dietary Breadth in the European Mid-Upper Paleolithic," *Proceedings of the National Academy of Sciences USA* 98

(2001): 6528–6532, doi:1073/Proceedings of the National Academy of Sciences USA.111155298; S. Kuhn and M. C. Stiner, "What's a Mother to Do? The Division of Labor among Neandertals and Modern Humans in Eurasia," *Current Anthropology* 47 (2006): 953–980.

9. H. Bocherens and D. Drucker, "Dietary Competition between Neanderthals and Modern Humans: Insights from Stable Isotopes," in *When Neanderthals and Modern Humans Met,* ed. N. Conard (Tübingen: Kerns Verlag, 2006), 129.

10. Ibid., 136.

11. H. Bocherens, "Diet and Ecology of Neanderthals: Implications from C and N Isotopes," in *Neanderthal Lifeways, Subsistence and Technology: One Hundred Fifty Years of Neanderthal Study,* eds. N. Conard and J. Richter (New York: Springer, 2011), 73–85.

12. V. Fabre, S. Condemi, A. Degioanni et al., "Neanderthals versus Modern Humans: Evidence for Resource Competition from Isotopic Modelling," *International Journal of Evolutionary Biology* 2011, Article ID 689315, doi:10.4061/2011/689315.

13. D. Salazar-García, R. Power, A. Sanchis Serra et al., "Neanderthal Diets in Central and Southeastern Mediterranean Iberia," *Quaternary International* 318 (2013): 3–18; D. Salazar-García, *Isótopos, dieta y movilidad en el País Valenciano: Aplicación a restos humanos del Paleolítico medio al Neolítico final* (Isotopes, Diet and Mobility in the Valencian Region; Application to Human Remains of the Middle Paleolithic and Final Neolithic) Ph.D. diss., Universitat de València, 2012.

14. Stringer et al., "Neanderthal Exploitation of Marine Mammals."

15. P. Shipman, "Separating 'Us' from 'Them': Neanderthal and Modern Human Behavior," *Proceedings of the National Academy of Sciences USA* 105 (2008): 14241–14242.

16. The observation was made to me by Curtis Marean in a personal communication, January 27, 2014.

17. C. Egelund, "Carcass Processing Intensity and Cutmark Creation: An Experimental Approach," *Plains Anthropologist* 48, no. 184 (2003): 39–51.

18. Kuhn and Stiner, "What's a Mother to Do?"

19. Higham et al., "Timing and Spatio-Temporal Patterning."

20. M. Germonpré, M. Udrescu, and E. Fiers, "Possible Evidence of Mammoth Hunting at the Neanderthal Site of Spy (Belgium)," *Quaternary International* (2012) 337: 28–42.

21. C. Kay, "Are Ecosystems Structured from the Top-Down or from the Bottom-Up?" in *Wilderness and Political Ecology: Aboriginal Influences and the Original State of Nature*, eds. C. Kay and R. Simmons (Salt Lake City: University of Utah Press, 2002), 215–237.

22. J. Wilczyński, P. Wojtal, and K. Sobczyk, "Spatial Organization of the Gravettian Mammoth Hunters' Site at Kraków Spadzista (Southern Poland)," *Journal of Archaeological Science* 39 (2012): 3627–3642; O. Soffer, *The Upper Paleolithic of the Central Russian Plain* (San Diego: Academic Press, 1985); R. Klein, *Ice-Age Hunters of the Ukraine* (Chicago: University of Chicago Press, 1973).

7. What Does an Invasion Look Like?

1. W. Ripple, J. Estes, R. Beschta et al., "Status and Ecological Effects of the World's Largest Carnivores," *Science* 343 (2014): 152, doi:10.1126/science.1241484.

2. E. Borer, B. Halpern, and E. Seabloom, "Asymmetry in Community Regulation: Effects of Predators and Productivity," *Ecology* 87 (2006): 2813–2820.

3. Ripple et al., "Status and Ecological Effects."

4. S. Fritz, R. Stephensen, R. Haynes et al., "Wolves and Humans," in *Wolves—Their Behavior, Ecology, and Conservation*, eds. D. Mech and L. Boitani (Chicago: University of Chicago Press, 2003), 294.

5. Kay, "Are Ecosystems Structured from the Top-Down?"

6. Data from D. Smith, R. Peterson, and D. Houston, "Yellowstone after Wolves," *BioScience* 53 (2003): 330–340.

7. J. W. Laundre, L. Hernandez, and W. J. Ripple, "The Landscape of Fear: Ecological Implications of Being Afraid," *Open Ecology Journal* 3 (2010): 1–7.

8. D. Smith and E. Bang, "Reintroduction of Wolves to Yellowstone National Park: History, Values and Ecosystem Restoration," in *Reintroduction of Top-Order Predators*, eds. M. Hayward and M. Somers (London: Blackwell, 2009), 92–124.

9. C. Wilmers, R. Crabtree, D. Smith et al., "Trophic Facilitation by Introduced Top Predators: Grey Wolf Subsidies to Scavengers in Yellowstone National Park," *Journal of Animal Ecology* 72 (2003): 909–916.

10. R. Crabtree and S. Sheldon, "The Ecological Role of Coyotes on Yellowstone's Northern Range," in *Carnivores in Ecosystems: The*

Yellowstone Experience, eds. T. W. Clark, A. P. Curlee, S. C. Minta, and P. M. Kareiva (New Haven, CT: Yale University Press, 1999), 127–163.

11. F. Palomares and T. Caro, "Interspecific Killing among Mammalian Carnivores," *American Naturalist* 153 (1999): 492–508.

12. Crabtree and Sheldon, "The Ecological Role of Coyotes."

13. D. Smith, R. Peterson, and D. Houston, "Yellowstone after Wolves."

14. C. Wilmers, R. Crabtree, D. Smith et al., "Resource Dispersion and Consumer Dominance: Scavenging at Wolf- and Hunter-Killed Carcasses in Greater Yellowstone, USA," *Ecology Letters* 6 (2003): 996–1003, doi:10.1046/j.1461-0248.2003.00522.x; Wilmers et al., "Trophic Facilitation."

15. P. Zarnetske, D. Skelly, and M. Urban, "Biotic Multipliers of Climate Change," *Science* 336 (2012): 1516–1518.

16. Various ranchers in Paradise Valley, personal communication to author, August 2, 2012; D. Mech, "Wolf Restoration to the Adirondacks and Advantages and Disadvantages of Public Participation in the Decisions," in *Wolves and Human Communities: Biology, Politics, and Ethics,* eds. V. Sharpe, B. Noron, and S. Donnelly (Washington, DC: Island Press, 2000), 13–22; V. Geist, "When Do Wolves Become Dangerous to Humans?," September 29, 2007, http://www.vargfakta.se/wp-content/uploads/2012/05/Geist-when-do -wolves-become-dangerous-to-humans-pt-1.pdf.

17. R. Wood, T. Higham, T. de Torres et al., "A New Date for the Neanderthals from El Sidrón Cave (Asturias, Northern Spain)," *Archaeometry* 55 (2013): 148–158, doi:10.1111/j.1475-4754.2012.00671.x; T. de Torres, J. Ortiz, R. Grün et al., "Dating of the Hominid (*Homo neanderthalensis*) Remains Accumulation from El Sidrón Cave (Borines, Asturias, North Spain): An Example of Multi-Methodological Approach to the Dating of Upper Pleistocene Sites," *Archaeometry* 52 (2010): 680–705, doi:10.1111/j/1475-4754.2009.00671.x.

18. J. Bermudez de Castro and P. Perez, "Enamel Hypoplasia in the Middle Pleistocene Hominids from Atapuerca (Spain)," *American Journal of Physical Anthropology* 96 (1995): 301–314; D. Guatelli-Steinberg, C. S. Larsen, and D. L. Hutchinson, "Prevalence and the Duration of Linear Enamel Hypoplasia: A Comparative Study of Neandertals and Inuit Foragers," *Journal of Human Evolution* 47 (2004): 65–84.

19. A. Rosas, E. Martínez, J. Cañaveras et al., "Paleobiology and Comparative Morphology of a Late Neanderthal Sample from El Sidrón, Asturias, Spain," *Proceedings of the National Academy of Sciences USA* 103 (2006): 19266–19271.
20. A. Defleur, O. Dutour, H. Valladas et al., "Cannibals among the Neanderthals?" *Nature* 362 (1993): 214.
21. A. Defleur, T. White, P. Valensi et al., "Neanderthal Cannibalism at Moula-Guercy, Ardèche, France," *Science* 286 (1999): 128–131.

8. Going, Going, Gone . . .

1. Bar-Yosef, "Eat What Is There," 338.
2. P. Mellars and J. French, "Tenfold Population Increase in Western Europe at the Neandertal-to-Modern Human Transition," *Science* 333 (2011): 623–627.
3. For example, see O. Bar-Yosef and J.-G. Bordes, "Who Were the Makers of the Châtelperronian Culture?" *Journal of Human Evolution* 59 (2010): 586–593; T. Higham, R. Jacobi, M. Julien et al., "Chronology of the Grotte du Renne (France) and Implications for the Context of Ornaments and Human Remains within the Châtelperronian," *Proceedings of the National Academy of Sciences USA* 107 (2010): 20234–20239.
4. P. Shipman, unpublished. By removing numbers based on Châtelperronian sites from their data sets, I gleaned the following figures: Numbers of sites: $N = 26$ Mousterian sites and 147 Aurignacian ones. Retouched tool densities/m^2/1,000 years: For nine Mousterian sites, the mean is 6.6 retouched tools/m^2/1,000 years versus twelve Aurignacian sites with a mean of 1.76/m^2/1,000 years. A two-tailed Mann-Whitney test shows this difference to be highly significant, with a value of $p < .002$. Meat weights: There are no meat weights for the Mousterian sites in their data set, so the value cannot be compared statistically to the Aurignacian mean figure of 152.8 kilograms for fifteen sites. Site area: The mean area of five Mousterian sites is 110 m^2 whereas that of twelve Aurignacian sites is 243.8 m^2. Although the Aurignacian sites are clearly larger in extent than the Mousterian ones, statistical evaluation is fairly meaningless because of the many factors (funding, size of crew, geographical factors) that influence how much of a site is excavated. Consult Mellars and French, "Tenfold Population Increase," for further information on procedures and comparisons.

5. B. Hockett and J. Haws, "Nutritional Ecology and the Human Demography of Neandertal Extinction," *Quaternary International* 137 (2005): 21–34, doi:10.1016/j.quaint.2004.11.017; B. Hockett, "The Consequence of Middle Paleolithic Diets on Pregnant Neandertal Women," *Quaternary International* 264 (2011): 78–82; L. Dalén, L. Orlando, B. Shapiro et al., "Partial Genetic Turnover in Neanderthals: Continuity in the East and Population Replacement in the West," *Molecular Biology and Evolution* 29 (2012): 1893–1897.

6. N. Conard, "The Demise of the Neanderthal Cultural Niche and the Beginning of the Upper Paleolithic in Southwestern Germany," in *Neanderthal Lifeways, Subsistence, and Technology: One Hundred Fifty Years of Neanderthal Study,* eds. N. Conard and J. Richter (New York: Springer, 2011), 228.

7. N. Conard, "Changing View of the Relationship between Neanderthals and Modern Humans," in *When Neanderthals and Modern Humans Met,* ed. N. Conard (Tübingen: Verlag, 2006), 11.

8. Higham et al., "The Timing and Spatio-Temporal Patterning."

9. D. Adler, O. Bar-Yosef, A. Belfer-Cohen et al., "Dating the Demise: Neanderthal Extinction and the Establishment of Modern Humans in the Southern Caucasus," *Journal of Human Evolution* 55 (2008): 817–833; Pinhasa et al., "Revised Age of Late Neanderthal Occupation"; J.-M. López-García, H.-A. Blain, M. Bennàsar et al., "Heinrich Event 4 Characterized by Terrestrial Proxies in South-western Europe," *Climate of the Past* 9 (2013): 1053–1064.

10. Higham et al., "Testing Models for the Beginnings of the Aurignacian."

11. C. Finlayson, F. Pacheco, J. Rodriguez-Vidal et al., "Late Survival of Neanderthals at the Southernmost Extreme of Europe," *Nature* 443 (2006): 850–853.

12. Dalén et al., "Partial Genetic Turnover in Neanderthals."

13. Defleur et al., "Neanderthal Cannibalism at Moula-Guercy"; Rosas et al., "Paleobiology and Comparative Morphology of a Late Neanderthal Sample."

14. S. Churchill, R. Franciscus, H. McKean-Peraza et al., "Shanidar 3 Neandertal Rib Puncture Wound and Paleolithic Weaponry," *Journal of Human Evolution* 57 (2009): 163–178, doi:10.1016/j.jhevol.2009.05.010.

15. F. Ramirez Rozzi, F. d'Errico, M. Vanhaeren et al., "Cutmarked Human Remains Bearing Neandertal Features and Modern Human

Remains Associated with the Aurignacian at Les Rois," *Journal of Anthropological Science* 87 (2009): 153–185.

16. C. Finlayson, *The Human Who Went Extinct; Why Neanderthals Died Out and We Survived* (New York: Oxford University Press, 2009), 118–119.

9. Guess Who Else Is Coming to Dinner?

1. Froehle and Churchill, "Energetic Competition between Neandertals and Anatomically Modern Humans"; L. Aiello and P. Wheeler, 2003. "Neanderthal Thermoregulation and the Glacial Climate," in *Neanderthals and Modern Humans in the European Landscape during the Last Glaciation,* eds. T. van Andel and W. Davies (Cambridge: McDonald Institute for Archaeological Research, University of Cambridge, 2003), 147–166; S. Churchill, "Bioenergetic Perspectives on Neandertal Thermoregulatory and Activity Budgets," in *Neanderthals Revisited: New Approaches and Perspectives,* eds. K. Harvati and T. Harrison (Dordrecht: Springer, 2006), 113–131; M. Sorensen and W. Leonard, "Neanderthal Energetics and Foraging Efficiency," *Journal of Human Evolution* 40 (2001): 483–495; A. Steegmann, F. Cerny, and T. Holliday, "Neandertal Cold Adaptation: Physiological and Energetic Factors," *American Journal of Human Biology* 14 (2002): 566–583.

2. Sorensen and Leonard, "Neanderthal Energetics and Foraging Efficiency"; Steegmann et al., "Neandertal Cold Adaptation."

3. Aiello and Wheeler, "Neanderthal Thermoregulation."

4. J. Gittleman and S. Thompson, "Energy Allocation in Mammalian Reproduction," *American Zoologist* 28 (1988): 863–875.

5. B. Hockett, "The Consequences of Middle Paleolithic Diets on Pregnant Neanderthal Women," *Quaternary International* 264 (2012): 78–82, p. 79.

6. Ibid., 81.

7. T. Holliday, "Postcranial Evidence of Cold Adaptation in European Neandertals," *American Journal of Physical Anthropology* 104 (1997): 245–258.

8. C. Carbone and J. Gittelman, "A Common Rule for the Scaling of Carnivore Density," *Science* 295 (2002): 2273–2276.

9. See excellent summary by S. E. Churchill, *Thin on the Ground* (New York: Basic Books, 2014).

10. C. Hertler and R. Volmer, "Assessing Prey Competition in Fossil Carnivore Communities—A Scenario for Prey Competition and Its

Evolutionary Consequences for Tigers in Pleistocene Java," *Palaeogeography, Palaeoclimatology, Palaeoecology* 257 (2008): 67–80; H. Hemmer, O. Owen-Smith, and M. Mills, "Predator–Prey Size Relationships in an African Large-Mammal Food Web," *Journal of Animal Ecology* 77 (2008): 173–183, doi:10.1111/j.1365-2656.2007.01314.x.

11. S. Münzel, M. Stiller, M. Hofreiter et al., "Pleistocene Bears in the Swabian Jura (Germany): Genetic Replacement, Ecological Displacement, Extinctions and Survival," *Quaternary International* 245 (2011): 225–237.

12. A. Turner and M. Antón, *The Big Cats and Their Fossil Relatives: An Illustrated Guide to Their Evolution and Natural History* (New York: Columbia University Press, 1997); W. Anyonge, "Body Mass in Large Extant and Extinct Carnivores," *Journal of Zoology London* 231 (1993): 339–350.

13. R. Guthrie, *Frozen Fauna of the Mammoth Steppe* (Chicago: University of Chicago Press, 1990).

14. Turner and Antón, *Big Cats.*

15. Anyonge, "Body Mass in Large Extant and Extinct Carnivores."

16. Churchill, *Thin on the Ground*, ch. 8.

17. Hemmer et al., "Predator–Prey Size Relationships."

18. Data from D. Brook and D. Bowman, "The Uncertain Blitzkrieg of Pleistocene Megafauna," *Journal of Biogeography* 31 (2004): 517–523.

19. C. Carbone, G. Mace, S. C. Roberts et al., "Energetic Constraints on the Diet of Terrestrial Carnivores," *Nature* 402 (1999): 287, doi:10.1038/46266.

20. Hertler and Volmer, "Assessing Prey Competition."

21. Churchill, *Thin on the Ground*, 271–273.

22. D. Stanford, R. Bonnichsen, and R. Morlan, "The Ginsberg Experiment: Modern and Prehistoric Evidence of a Bone-Flaking Technology," *Science* 212 (1981): 438–440, doi:10.1126/science .212.4493.438.

23. D. Grayson and F. Delpech, "The Large Mammals of Roc de Combe (Lot, France): The Châtelperronian and Aurignacian Assemblages," *Journal of Anthropological Archaeology* 27 (2008): 359.

24. H. Bocherens, D. Drucker, D. Bonjean et al., "Isotopic Evidence for Dietary Ecology of Cave Lion (*Panthera spelaea*) in North-Western Europe: Prey Choice, Competition and Implications for Extinction," *Quaternary International* 245 (2011): 249–261.

25. Churchill, *Thin on the Ground*, 264–276.

26. N. Carter, B. Shrestha, J. Karki, N. Pradhan et al., "Coexistence between Wildlife and Humans at Fine Spatial Scales," *Proceedings of the National Academy of Sciences USA* 109 (2012): 15360–15365.

27. E. Ghezzo, A. Palchetti, and L. Rook, "Recovering Data from Historical Collections: Stratigraphic and Spatial Reconstruction of the Outstanding Carnivoran Record from the Late Pleistocene Equi Cave (Apuane Alps, Italy)," *Quaternary Science Reviews* 96 (2014): 168–179.

28. C. Diedrich, "Late Pleistocene Leopards across Europe: Northernmost European German Population, Highest Elevated Records in the Swiss Alps, Complete Skeletons in the Bosnia Herzegowina Dinarids and Comparison to the Ice Age Cave Art," *Quaternary Science Reviews* 76 (2013): 167–193; G. von Petzinger and A. Nowell, "A Place in Time: Situating Chauvet within the Long Chronology of Symbolic Behavioral Development," *Journal of Human Evolution*, in press, doi:10.1016/j.jhevol.2014.02.022.

10. Bearing Up under Competition Pressure

1. Münzel et al., "Pleistocene Bears in the Swabian Jura."

2. Ibid., 231.

3. "True Causes for Extinction of Cave Bear Revealed: More Human Expansion than Climate Change," *Science News*, August 25, 2010.

4. M. Stiller, G. Baryshnikov, H. Bocherens et al., "Withering Away—25,000 Years of Genetic Decline Preceded Cave Bear Extinction," *Molecular Biology and Evolution* 27 (2010): 975–978.

5. Mellars and French, "Tenfold Population Increase."

6. Münzel et al., "Pleistocene Bears in the Swabian Jura," 232–233.

7. Froehle and Churchill, "Energetic Competition between Neandertals and Anatomically Modern Humans."

8. P. Wojtal, J. Wilczyński, Z. Bocheński et al., "The Scene of Spectacular Feasts: Animal Remains from Pavlov I South-east, Czech Republic," *Quaternary International* 252 (2012): 122–141.

9. S. Kuhn and M. C. Stiner, "What's a Mother to Do? The Division of Labor among Neandertals and Modern Humans in Eurasia," *Current Anthropology* 47 (2006): 953–980.

10. G. Haynes, *Mammoths, Mastodonts, and Elephants: Biology, Behavior, and the Fossil Record* (Cambridge: Cambridge University Press, 1991).

11. E. Wing and A. Brown, *Paleonutrition* (New York: Academic Press, 1979).

12. Hortolà and Martínez-Navarro, "Quaternary Megafaunal Extinction and the Fate of Neanderthals."

11. The Jagger Principle

1. K. Richards and M. Jagger, "You Can't Always Get What You Want." *Let It Bleed*, ABKCO Records, 2002.

2. Mellars and French, "Tenfold Population Increase."

3. N. Conard, "The Cultural Niche"; N. Conard, M. Bolus, P. Goldberg et al., "The Last Neanderthals and the First Modern Humans in the Swabian Jura," in *When Neanderthals and Modern Humans Met*, ed. N. Conard (Tübingen: Kerns Verlag, 2006), 305–342.

4. Density data are from Haynes, *Mammoths, Mastodonts, and Elephants*; G. Haynes, "Mammoth Landscapes: Good Country for Hunter-Gatherers," *Quaternary International* 142/143 (2006): 20–29; F. De Boer, F. van Langevelde, H. Prins et al., "Understanding Spatial Differences in African Elephant Densities and Occurrence, a Continent-Wide Analysis," *Biological Conservation* 159 (2013): 468–476; IUCN African Elephant Specialist Group, www.elephant database.org.

5. Wilczyński et al., "Spatial Organization of the Gravettian Mammoth Hunters' Site," 3638.

6. L. Demay, S. Péan, and M. Patou-Mathis, "Mammoths used as food and building resources by Neanderthals: Zooarchaeological study applied to layer 4, Molodova I (Ukraine)," *Quaternary International* 276/277 (2012): 212–226.

7. R. H. Gargett, "One Mammoth Steppe Too Far," *Subversive Archaeologist* (blog), December 21, 2011, http://www.thesubver sivearchaeologist.com/2011/12/one-mammoth-steppe-too-far.html.

8. Higham et al., "The Timing and Spatio-Temporal Patterning."

9. H. Bocherens and D. Drucker, "Dietary Competition between Neanderthals and Modern Humans: Insights from Stable Isotopes," in *When Neanderthals and Modern Humans Met*, ed. N. Conard (Tübingen: Kerns Verlag, 2006), 129–143; Germonpré et al., "Possible Evidence of Mammoth Hunting"; P. Semal, H. Rougier, I. Crevecoeur et al., "New Data on the Late Neandertals: Direct Dating of the Belgian Spy Fossils," *American Journal of Physical Anthropology* 138 (2009): 421–428.

10. N. McDermott, "Did Climate Change Drive the Woolly Mammoth to Extinction? Genetic Tests Reveal Species Declined as Weather Warmed," *Daily Mail*, September 11, 2013.

12. Dogged

1. M. Germonpré, M. Sablin, R. E. Stevens et al., "Fossil Dogs and Wolves from Palaeolithic Sites in Belgium, the Ukraine and Russia: Osteometry, Ancient DNA and Stable Isotopes," *Journal of Archaeological Science* 36 (2009): 473–490.

2. M. Germonpré, M. Lázničková-Galetová, and M. Sablin, "Palaeolithic Dog Skulls at the Gravettian Předmostí Site, the Czech Republic," *Journal of Archaeological Science* 39 (2012): 184–202.

3. M. Germonpré, J. Räikönnen, M. Lázničková-Galetová et al., "Mandibles from Palaeolithic Dogs and Pleistocene Wolves at Předmostí, the Czech Republic," in *The World of Gravettian Hunters,* ed. P. Wojtal, Institute of Systematics and the Evolution of Animals, Polish Academy of Sciences (2013), 23–24; M. Germonpré, M. Lázničková-Galetová, R. Losey et al., "Large Canids at the Gravettian Předmostí site, the Czech Republic: The Mandible," *Quaternary International* (in press): 1–19.

4. Germonpré et al., "Fossil Dogs and Wolves," 482.

5. N. Ovodov, S. Crockford, Y. Kuzmina et al., "A 33,000-Year-Old Incipient Dog from the Altai Mountains of Siberia: Evidence of the Earliest Domestication Disrupted by the Last Glacial Maximum," *PLoS ONE* 6, no. 7 (2011): e22821.

6. O. Thalmann, B. Shapiro, P. Cui et al., "Complete Mitochondrial Genomes of Ancient Canids Suggest a European Origin of Domestic Dogs," *Science* 342 (2013): 871–874.

7. Ibid., 872.

8. R. Wayne, personal communication to author, 2012.

9. J. Avise, *On Evolution* (Baltimore: Johns Hopkins University Press, 2007), 49–50.

10. E. Ostrander and R. Wayne, "The Canine Genome," *Genome Research* 15 (2005): 1706–1716, doi:10.1101/gr.3736605.

11. R. Wayne, "Cranial Morphology of Domestic and Wild Canids: The Influence of Development on Morphological Change," *Evolution* 40 (1986): 243–261; R. Wayne, "Limb Morphology of Domestic and Wild Canids: The Influence of Development on Morphologic Change," *Journal of Morphology* 187 (1986): 301–319.

12. See the review in J. Clutton-Brock, *Animals as Domesticates: A World View through History* (East Lansing: Michigan State University Press, 2012).

13. R. Coppinger and L. Coppinger, *Dogs: A Startling New Under-standing of Canine Origin, Behavior, and Evolution* (New York: Scribner, 2001).

14. V. Geist, "When Do Wolves Become Dangerous to Humans?," September 29, 2007, http://www.vargfakta.se/wp-content/uploads /2012/05/Geist-when-do-wolves-become-dangerous-to-humans-pt -1.pdf.

15. O. Soffer, personal communication to author, March 15, 2013.

16. L. Marquer, V. Lebretona, T. Otto et al., "Charcoal Scarcity in Epigravettian Settlements with Mammoth Bone Dwellings: The Taphonomic Evidence from Mezhyrich (Ukraine)," *Journal of Archaeological Science* 39 (2012): 109–120.

17. See, e.g., S. Leshchinshy and O. Bukharova, "Geochemical Stress of the Kraków-Spadzista Street Mammoth Population Demonstrated by Electron Microscopy," in *The World of Gravettian Hunters, Institute of Systematics and the Evolution of Animals,* ed. P. Wojtal (Kraków: Polish Academy of Sciences, 2013), 45–49; S. Leshchinsky, "Lugovs-koye: Environment, Taphonomy, and Origin of a Paleofaunal Site," *Archaeology, Ethnology & Anthropology of Eurasia* 1, no. 25 (2006): 33–40, doi:10.1134/S1563011006010026.

18. N. Bicho, A. Pastoors, and B. Auffermann, eds., *Human's Best Friends—Dogs . . . and Fire! Pleistocene Foragers on the Iberian Peninsula: Their Culture and Environment* (Mettmann: Wissen-schaftliche Schriften des Neanderthal Museums 7, 2013): 217–242.

19. V. Ruusila and M. Pesonen, "Interspecific Benefits in Human (*Homo sapiens*) Hunting: Benefits of a Barking Dog," *Annals of the Zoologica Fennici* 41 (2004): 545–549.

20. J. Koster and K. Tankersley, "Heterogeneity of Hunting Ability and Nutritional Status among Domestic Dogs in Lowland Nicaragua, Hunting Ability," *Proceedings of the National Academy of Sciences USA* 109 (2012): 463–470, doi:10.1073/pnas.1112515109.

21. K. Lupo, "A Dog Is for Hunting," in *Ethnozooarchaeology,* eds. U. Albarella and A. Trentacoste (Oxford: Oxbow Press, 2011), 4–12.

22. M. Stiner and S. Kuhn, "Paleolithic Diet and the Division of Labor in Mediterranean Eurasia," in *The Evolution of Hominin Diets; Integrating Approaches to the Study of Paleolithic Subsistence,* eds. J.-J. Hublin and M. P. Richards (Springer Science and Business Media B.V., 2009): 157–169, p. 161.

23. C. Arnold, "Possible Evidence of Domestic Dog in a Paleoeskimo Context," *Arctic* 32 (1979): 263.
24. J. Speth, K. Newlander, A. White et al., "Early Paleoindian Big-Game Hunting in North America: Provisioning or Politics?" *Quaternary International* 285 (2013): 121.
25. C. Turner, "Teeth, Needles, Dogs and Siberia: Bioarchaeological Evidence for the Colonization of the New World," in *The First Americans: The Pleistocene Colonization of the New World*, ed. N. Jablonski (San Francisco: California Academy of Sciences, 2002), 123–158.
26. S. Fiedel, "Man's Best Friend and Mammoth's Worst Enemy? A Speculative Essay on the Role of Dogs in Paleoindian Colonization and Megafaunal Extinction," *World Archaeology* 37 (2005): 11–25.
27. D. Morey and K. Aaris-Sørensen, "Paleoeskimo Dogs of the Eastern Arctic," *Arctic* 55 (2002): 44–56.
28. "Hunting a Polar Bear with Dogs," October 18, 2012, http://retriever man.net/2012/10/18/hunting-a-polar-bear-with-dogs/; S. Vilhjalmur, *The Friendly Arctic: The Story of Five Years in Polar Regions* (New York: Macmillan: 1921).
29. Kuhn and Stiner, "What's a Mother to Do?"
30. A. Vanak and M. Gompper, "Dogs *Canis familiaris* as Carnivores: Their Role and Function in Intraguild Competition," *Mammalian Review* (2009) 148: 265–283, 281.
31. Wojtal et al., "The Scene of Spectacular Feasts," 135–137.
32. Data are from R. Musil, "Palaeoenvironment at Gravettian Sites in Central Europe with Emphasis on Moravia (Czech Republic)," *Quartär* 57 (2010): 95–123.
33. H. Bocherens, D. Drucker, M. Germonpré et al., "Reconstruction of the Gravettian Food-Web at Předmostí I Using Multi-Isotopic Tracking (^{13}C, ^{15}N, ^{34}S) of Bone Collagen," *Quaternary International*, in press.

13. Why Dogs?

1. F. S. Galton, "The First Steps Towards the Domestication of Animals," *Transactions of the Ethnological Society, London* (1863) n.s. 1: 122–138; J. Clutton-Brock, *A Natural History of Domesticated Animals*, 2nd edition (Cambridge: Cambridge University Press, 1999).
2. A. Sherratt, "The Secondary Exploitation of Animals in the Old World," *World Archaeology, Transhumance and Pastoralism* 15, no. 1

(1983): 90–104; A. Sherratt, "Plough and Pastoralism: Aspects of the Secondary Products Revolution," in *Pattern of the Past: Studies in Honour of David Clarke*, eds. I. Hodder, G. Isaac, and N. Hammond (Cambridge: Cambridge University Press, 1981): 261–305.

3. P. Shipman, "And the Last Shall be First," in *Animal Secondary Products: Archaeological Perspectives on Domestic Animal Exploitation in the Neolithic and Bronze Age*, ed. H. Greenfield (Oxford: Oxbow Books, 2014), 40–54.

4. L. Lord, "A Comparison of the Sensory Development of Wolves *(Canis lupus lupus)* and Dogs *(Canis lupus familiaris),*" *Ethology* 119 (2013): 110–120, doi:10.1111/eth.12044.

5. B. Hare and V. Woods, *The Genius of Dogs: How Dogs Are Smarter than You Think* (New York: Dutton, 2013), 76–77.

6. L. Trut, "Early Canid Domestication: The Farm-Fox Experiment," *American Scientist* 87 (1999): 160, doi:10.1511/1999.2.160; L. Trut, E. Naumenko, and D. Belyaev, "Change in Pituitary-Adrenal Function in Silver Foxes under Selection for Domestication," *Genetika* 5 (1972): 35–43 (in Russian, English abstract).

7. Shipman, *The Animal Connection;* Shipman, "And the Last."

8. J. Bohannon, "Who's (Socially) Smarter: The Dog or the Wolf?" *Science Now,* May 28, 2013.

9. P. Smith and C. Litchfield, "How Well Do Dingoes, *Canis dingo,* Perform on the Detour Task?" *Animal Behaviour* 80 (2010): 155–162.

10. Geist, "When Do Wolves Become Dangerous?"

11. Hare and Woods, *Genius of Dogs,* 14.

12. Soffer, *Upper Paleolithic of the Central Russian Plain,* 258, 187.

13. Palomares and Caro, "Interspecific Killing."

14. Vanak and Gompper, "Dogs *Canis familiaris* as Carnivores."

15. Wojtal et al., "The Scene of Spectacular Feasts."

16. G. Haynes, "Utilization and Skeletal Disturbances of North American Prey Carcasses," *Arctic* 35 (1982): 266–382.

17. M. Stiner, "Comparative Ecology and Taphonomy of Spotted Hyenas, Humans, and Wolves in Pleistocene Italy," *Revue de Paléobiologie, Génève* 23 (2004): 771–785.

18. D. Morey, "Burying Key Evidence: The Social Bond between Dogs and People," *Journal of Archaeological Science* 33 (2006): 159.

19. R. Losey, S. Garvie-Lok, J. Leonard et al., "Burying Dogs in Ancient Cis-Baikal, Siberia: Temporal Trends and Relationships with Human Diet and Subsistence Practices," *PLoS ONE* 8, no. 5 (2013): 63740–63763; R. Losey, V. Bazaliski, S. Garvie-Lok et al.,

"Canids as Persons: Early Neolithic Dog and Wolf Burials, Cis-Baikal, Siberia," *Journal of Anthropological Archaeology* 30 (2011): 174–189.

20. K. Maška, "Maška's Diary: The Text of Maška's Diary," in *Early Modern Humans from Předmostí Near Přerova, Czech Republic: A New Reading of Old Documentation*, eds. J. Velemínská and J. Brůžek (Prague: Academia, 2008), 181–188 (English translation).

21. Germonpré et al., "Mandibles from Palaeolithic Dogs."

22. V. Van Valkenburgh and F. Hertel, "Tough Times at La Brea: Tooth Breakage in Large Carnivores of the Late Pleistocene," *Science* 261 (1993): 456–459.

23. R. Losey, E. Jessup, T. Nomokonova et al., "Craniomandibular Trauma and Tooth Loss in Northern Dogs and Wolves: Implications for the Archaeological Study of Dog Husbandry and Domestication," *PLoS ONE* 9, no. 6 (2014): e99746, doi:10.1371/journal.pone.0099746.

24. R. White, "Systems of Personal Ornamentation in the Early Upper Palaeolithic: Methodological Challenges and New Observations," in *Rethinking the Human Revolution: New Behavioural and Biological Perspectives on the Origin and Dispersal of Modern Humans*, eds. P. Mellars, K. Boyle, O. Bar-Yosef, and C. Stringer (Cambridge: McDonald Institute for Archaeological Research, University of Cambridge, 2007), 299.

25. P. Bahn, personal communication to author, 2011.

26. A. Pike-Tay, personal communication to author, 2011.

27. K. Sterelny, *The Evolved Apprentice: How Evolution Made Humans Unique* (Cambridge, MA: MIT Press, 2012).

14. When Is a Wolf Not a Wolf?

1. D. Morey, "Burying Key Evidence: The Social Bond between Dogs and People," *Journal of Archaeological Science* 33 (2006): 167.

2. Lord, "Comparison of the Sensory Development of Wolves."

3. P. Shipman, "Do the Eyes Have It?" *American Scientist* 100 (2012): 198–201.

4. H. Kobayashi and S. Kohshima, "Unique Morphology of the Human Eye and Its Adaptive Meaning: Comparative Studies on External Morphology of the Primate Eye," *Journal of Human Evolution* 40 (2001): 419–435.

5. S. Ueda, G. Kumagi, Y. Otaki et al., "A Comparison of Facial Color Pattern and Gazing Behavior in Canid Species Suggests Gaze

Communication in Gray Wolves (*Canis lupus*)," *PLoS ONE* 9, no. 6 (2014): 98217–98223, doi:10.1371/journal.pone.0098217.

6. M. Tomasello, B. Hare, H. Lehmann et al., "Reliance on Head Versus Eyes in the Gaze Following of Great Apes and Human Infants: The Cooperative Eye Hypothesis," *Journal of Human Evolution* 52 (2007): 314–320.

7. Anne Pusey, personal communication to author, Jan. 27, 2012.

8. F. Range and Z. Viránya, "Development of Gaze Following Abilities in Wolves (*Canis lupus*)," *PLoS ONE* 6, no. 2 (2012): e16888, doi:10.1371/journal.pone.0016888.

9. P. Pongrácz, Á. Miklósi, K. Timár-Geng et al., "Verbal Attention Getting as a Key Factor in Social Learning between Dog (*Canis familiaris*) and Human," *Journal of Comparative Psychology* 118 (2004): 375–383; B. Hare and M. Tomasello, "Human-Like Social Skills in Dogs?" *Trends in Cognitive Sciences* 9 (2005): 439–444; B. Hare and M. Tomasello, "Domestic Dogs (*Canis familiaris*) Use Human and Conspecific Social Cues to Locate Hidden Food," *Journal of Comparative Psychology* 113 (1999): 173–177; K. Soproni, Á. Miklósi, J. Topál et al., "Dogs' (*Canis familiaris*) Responsiveness to Human Pointing Gestures," *Journal of Comparative Psychology* 116 (2002): 27–34; Á. Miklósi, E. Kubinyi, J. Topál et al., "A Simple Reason for a Big Difference: Wolves Do Not Look Back at Humans but Dogs Do," *Current Biology* 13 (2003): 763–766.

10. D. Nogués-Bravo, R. Ohlemüller, P. Batra et al., "Climate Predictors of Late Quaternary Extinctions," *Evolution* 64 (2010): 2442–2449, doi:10.1111/j.1558-5646.2010.01009.x.

15. What Happened and Why

1. Dalén et al., "Partial Genetic Turnover in Neanderthals."

2. J. Reumer, L. Rook, K. Van Der Borg et al., "Late Pleistocene Survival of the Saber-Toothed Cat *Homotherium* in Northwestern Europe." *Journal of Vertebrate Paleontology* 23 (2003): 260, doi:10.1671/0272-4634(2003)23[260:LPSOTS]2.0.CO;2.

3. L. Rook, personal communication to author, July 31, 2014.

4. A. Stuart and A. Lister, "New Radiocarbon Evidence on the Extirpation of the Spotted Hyaena (*Crocuta crocuta* (Erxl.)) in Northern Eurasia," *Quaternary Science Reviews* 96 (2014): 108–116; H. Bocherens, M. Stiller, K. A. Hobson et al., "Niche Partition between Two Sympatric Genetically Distinct Bears from Austria *(Ursus spelaeus* and *Ursus ingressus)*," *Quaternary International* 245

(2011): 238–248; A. Stuart and A. Lister, "Extinction Chronology of the Cave Lion *Panthera spelaea*," *Quaternary Science Reviews* 30 (2011): 2329–2340; Münzel et al., "Pleistocene Bears"; Stiller et al., "Withering Away"; R. Barnett, B. Shapiro, I. Barnes et al., "Phylogeography of Lions (*Panthera leo* ssp.) Reveals Three Distinct Taxa and a Late Pleistocene Reduction in Genetic Diversity," *Molecular Ecology* 18 (2009): 1668–1677, doi:10.1111/j. 1365-294X.2009.04134.x; M. Hofreiter, G. Rabedder, V. Jaenicke-Després et al., "Evidence for Reproductive Isolation between Cave Bear Populations," *Current Biology* 14 (2004): 40–43; M. Hofreiter, Ch. Capelli, M. Krings et al., "Ancient DNA Analyses Reveal High Mitochondrial DNA Sequences Diversity and Parallel Morphological Evolution of Late Pleistocene Cave Bears," *Molecular Biology and Evolution* 19 (2002): 1244–1250.

5. G. Prescott, D. Williams, A. Balmford et al., "Quantitative Global Analysis of the Role of Climate and People in Explaining Late Quaternary Megafaunal Extinctions," *Proceedings of the National Academy of Sciences USA* 109, no. 12 (2012): 4527–4531.

6. Nogués-Bravo et al., "Climate Predictors of Late Quaternary Extinctions."

CREDITS

increase in western Europe at the Neandertal-to-modern human transition." *Science* (July 29, 2011), figure 3. Reprinted with permission from AAAS.

Figure 8.2. A sterile layer at Geissenklösterle. From Nicholas J. Conard, "The demise of the Neanderthal cultural niche and the beginning of the Upper Paleolithic in southwestern Germany," in *Neanderthal Lifeways, Subsistence and Technology*, eds. Nicholas Conard and Jürgen Richter (New York: Springer), figure 19.4. Copyright © 2011 University of Tübingen.

Figure 9.1. Body size influences size of preferred prey. Copyright © 2014 Jeffrey Mathison.

Figure 10.1. Flint projectile point in cave bear vertebra. From Susanne C. Münzel and Nicholas J. Conard, "Cave bear hunting in Hohle Fels Cave in the Ach Valley of the Swabian Jura." *Revue de Paléobiologie* 23(2):877–885, figure 10. Copyright © 2004 University of Tübingen.

Figure 10.2. Close-up, bear vertebra. From Susanne C. Münzel and Nicholas J. Conard, "Cave bear hunting in Hohle Fels Cave in the Ach Valley of the Swabian Jura." *Revue de Paléobiologie* 23(2):877–885, figure 11. Copyright © 2004 University of Tübingen.

Figure 10.3. Bear remains in Hohle Fells. From Susanne C. Münzel, Mathias Stiller, Michael Hofreiter, Alissa Mittnik, Nicholas J. Conard, and Hervé Bocherens, "Pleistocene bears in the Swabian Jura (Germany): Genetic replacement, ecological displacement, extinctions and survival." *Quaternary International* 245:225–237, figure 5. Copyright © 2011 University of Tübingen.

Figure 11.1. Mammoth bones at Kraków-Spadzista Street excavation. Copyright © 2014 Piotr Wojtal. Reprinted with permission.

Figure 11.2. Locations of sites with mammoth bone huts. Copyright © 2014 Jeffrey Mathison.

Figure 11.3. Mammoth ranges over time. Map after E. Palkopoulou et al. (2013), Holarctic genetic structure and range dynamics in the woolly mammoth. *Proceedings of the Royal Society B* 280:20131910. Copyright © 2014 Jeffrey Mathison.

Figure 12.1. Skull of Paleolithic dog, Goyet Cave. Reprinted from Mietje Germonpré, Mikhail V. Sablin, Rhiannon E. Stevens, Robert E. M. Hedges, Michael Hofreiter, Mathias Stiller, and Viviane R. Després, "Fossil dogs and wolves from Palaeolithic sites in Belgium, the Ukraine and Russia." *Journal of Archaeological Science* 36 (2009):473–490, figure 11, with permission from Elsevier.

Figure 12.2. Genetic relationships among fossil and living canids. From O. Thalmann et al., "Complete mitochondrial genomes of ancient canids suggest a European origin of domestic dogs." *Science* (November 15, 2013), 871–874, figure 1. Reprinted with permission from AAAS.

Figure 12.3. Wolves hunting bison, Yellowstone National Park. The National Park Service.

Figure 12.4. Clay fragments from Dolní Věstonice and Pavlov. Copyright © 2013 Pat Shipman, courtesy Dolní Věstonice Museum.

Figure 13.1. Dog from Předmostí, mortuary ritual. Copyright © 2014 Mietje Germonpré. Reprinted with permission. Courtesy of the Moravian Museum, Brno, the Czech Republic.

Figure 14.1. Three types of canid faces. From Sayoko Ueda, Gaku Kumagai, Yusuke Otaki, Shinya Yamaguchi, and Shiro Kohshima, "A comparison of facial color pattern and gazing behavior in canid species suggests gaze communication in gray wolves (*Canis lupus*)." *PLoS One* 9(6): e98217, figure 2. Copyright © 2014 Ueda et al. Reprinted with permission.

Figure 14.2. Early modern humans hunting mammoth with wolf-dogs. Copyright © 2014 Dan Burr.

INDEX

DNA *(continued)*
nuclear, 22–25, 175–180; of
Neanderthals, 23–30, 114–115
Dog (*Canis familiaris*): as hunting aid,
184–186; behavior of vs. wolves
(*Canis lupus*), 180; diet compared to
wolves at Předmostí, 192–193;
domestication of, 168, 180–182,
194–196; genome of, 172–179; in
guarding camp, 188; incipient dog,
172; in hauling, 187–188; lack of
depiction in prehistoric art,
211–212; mortuary treatment of,
207–210; reaction to wolves,
189–190; reaction to humans,
220–221; use in Arctic, 187–188
Domestication, 4–5, 197–198,
228–231; silver fox experiment,
198–201; species' suitability,
194–195
Drucker, Dorothée, 75–76

Elton, Charles, 19, 70
Eriksson, Anders, 27
Estes, James, 67
Extinction, causes of, 17–19, 63–65
Eyes, 215–219

Fiber, use of, 191–192
Finlayson, Clive, 37, 41, 56, 61–63, 65,
67, 78, 113, 117–119, 135, 140
French, Jennifer, 107–112, 148
Fur, use of by modern humans,
148–150, 154–155

García-Berthou, Emili, 18
Geist, Valerius, 180, 202
Germonpré, Mietje, 83, 167–179, 187,
192, 207–209
Gittelman, John, 125

Global Invasive Species Database, 1–2
Grayson, Don, 137–138
Green, Richard, 26
Grinnell, Joseph, 71
Guild, 71–72, 125; intraguild violence,
84, 87–88, 95, 99–101, 103–104,
106–107, 115–116, 203, 209,
223–224; predatory, 72, 125–130,
136, 141–142, 166, 181, 227, 231

Hare, Brian, 199–202
Heinrich Event, 52
Hertler, Christine, 131–133
Higham, Thomas, 37–43
Hockett, Brian, 123–124
Holt, Robert, 67
Homo sapiens: abilities, 31, 46–47,
56–57, 72; as predators, 48; role in
extinctions, 2–6, 19; vs. *Homo
neanderthalensis*, 20

Iberian refugia, 37, 39–41, 77–80,
117–118
International Union for the Conservation of Nature (IUCN), 1, 18,
63–64
Invasive species, 16–19, 45–46

Jagger Principle, 156

Kay, Charles, 83–84, 90
Kobayashi, Hiromo, 215
Koster, Jeremy, 186
Krings, M., 23
K selection, 68–69
Kuhn, Steven, 80, 152, 186–189

Leshchinsky, S. V., 183–184
Levant, 26, 50, 53–59
Lister, Adrian, 163, 165